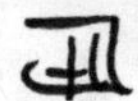

# Curve and Surface Design

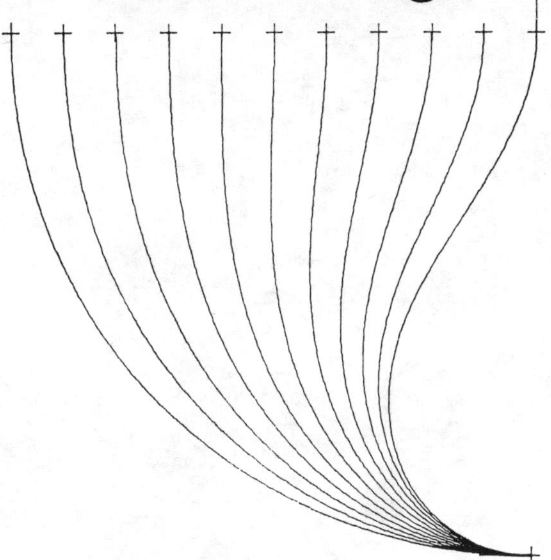

# Curve and Surface Design

## Edited by
## Hans Hagen
## Universität Kaiserslautern

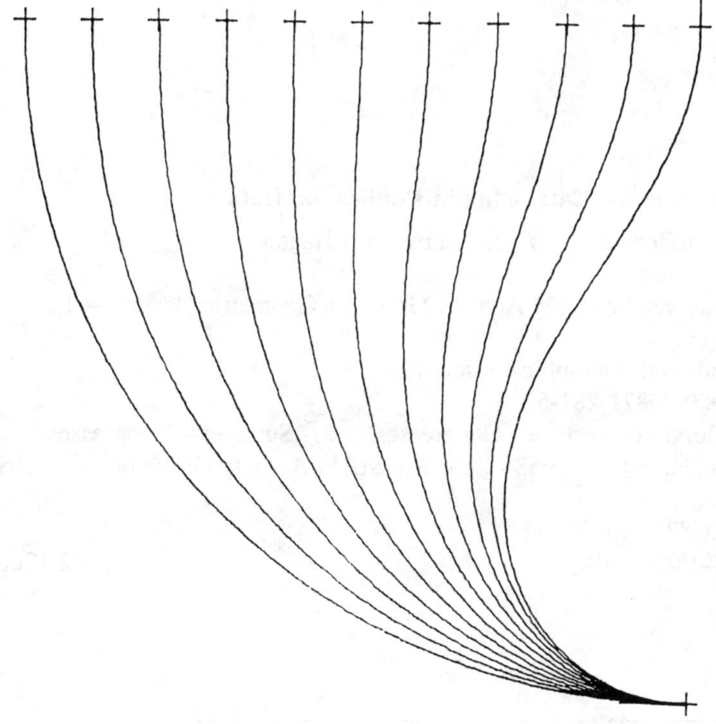

siam.

Philadelphia

Society for Industrial and Applied Mathematics

# Geometric Design Publications

Editor
Gerald E. Farin
Arizona State University

Farin, Gerald E., editor, Geometric Modeling: Algorithms and New Trends (1987)
Farin, Gerald E., editor, NURBS for Curve and Surface Design (1991)
Barnhill, Robert E., editor, Geometry Processing for Design and Manufacturing (1992)
Hagen, Hans, editor, Curve and Surface Design (1992)
Hagen, Hans, editor, Topics in Surface Modeling (1992)

**Library of Congress Cataloging-in-Publication Data**

Curve and surface design  /  edited by Hans Hagen.
    p.  cm.
    "Sponsored by SIAM Activity Group on Geometric Design"—T. p. verso.
    Includes bibliographical references.
    ISBN 0-89871-281-5
    1. Curves on surface—Congresses.   2. Surfaces—Congresses.
I. Hagen, H. (Hans), 1953–  .   II. SIAM Activity Group on Geometric Design.
QA643.C87   1992
516'.6—dc20                       92-12066

Sponsored by SIAM Activity Group on Geometric Design.

# Preface

Curve and surface design in industrial applications still has many unresolved issues, even though research in this field has been ongoing for more than 30 years. This book is a collection of new ideas and results on topics of curve and surface design. It is intended not only for research in the academic environment, but for practical use in industrial applications.

The starting point for this book was the SIAM Conference on Geometric Design held in Tempe, Arizona, from November 6–10, 1989. Many of the ideas contained in this volume were presented there. Additionally, some of the leading experts in the field were invited to contribute. About 60% of the submissions were finally accepted for publication.

The book is organized in two sections: curve design and nontensor product surfaces.

*Part 1. Curve Design.* This section contains the newest developments in curve design. The first two contributions are concerned with minimal-energy splines. Smoothing algorithms are frequently based on the idea of simulating the behavior of elastic materials that tend to minimize their elastic energy. The energy stored in a thin elastic beam is proportional to the integral of the squared curvature of the beam with the length of the beam as the variable of integration. Curves minimizing this energy functional are therefore used to define a model of a smooth curve shape.

The article by Brunnett presents an analysis of the criterion of minimal energy, giving new insight into the properties of minimal-energy splines and their behavior. The second paper on this topic, by Jou and Han, shows how to introduce various end constraints in the concept of minimal-energy splines.

A milestone in free-form curve design was Nielson's development of a piecewise polynomial alternative to splines under tension, the so-called ν-spline, a piecewise cubic, curvature continuous spline. Hagen generalized Nielson's approach resulting in polynomials of degree $n = 2L - 1$, $L > 2$. This concept of geometric spline curves includes Nielson's ν-splines for $L = 2$ and curvature and torsion continuous quintics for $L = 3$, the so-called τ-splines. Geometric spline curves provide shape parameters having the characteristics of point weights.

The idea of interval weights is attributed to Salkauskas. In many applications there is a need to combine both concepts. Just recently, Foley generalized Nielson's ν-spline to an interval-weighted ν-spline. The gap to τ-splines is bridged in this volume. Lasser and Hagen present interval-weighted τ-splines, and Neuser presents curve and surface interpolating techniques using quintic weighted τ-splines.

The latest progress in weighted spline methods is presented by Bos and Salkauskas. The topics are completed by Eck's algorithms for geometric spline curves. The contribution of Fritsch and Nielson is concerned with the problem of determining the distance between parametric curves. They present an elegant, application-oriented solution.

*Part 2. Nontensor Product Surfaces.* The problem of fitting a surface through a set of three-dimensional data points is a key problem in the field of computer aided geometric design. Tensor product schemes work quite well for modeling surfaces based on strict rectangular networks, but very often applications need more general topologies.

In this section, we present three different approaches dealing with this problem. DeRose et al. give a critical survey on triangular surface interpolating methods. The paper by Bloor and Wilson describes a method for surface generation, whereby a smooth surface is generated as a solution to a boundary value problem. In his contribution, Dæhlen considers bivariate box splines with special emphasis on three-, five-, and six-sided surface design.

I would like to thank Robert E. Barnhill, Gerald Farin, and the SIAM staff for all their effort in organizing the SIAM conference, and especially for their kind help and advice in initiating this book. I would like to express my appreciation for the referees—their support and contribution are always important, but rarely acknowledged. Their names are listed below. Thanks also go to Dieter Lasser, Guido Brunnett, and Ernst Gschwind for numerous and valuable discussions. Last but not least, this book would not have been possible without the countless hours spent by our secretaries Elisabeth Gruys and Karen Lasser on the paperwork that always seems to accompany such as enterprise.

Hans Hagen<br>Universität Kaiserslautern

## Referees

| | | |
|---|---|---|
| G. Brunnett | G. Farin | G. Nielson |
| W. Dankwort | E. Gschwind | D. Roller |
| T. DeRose | H. Hagen | P. Santarelli |
| R. Franke | J. Hoschek | |
| T. Foley | D. Lasser | |

# List of Contributors

**M. I. G. Bloor**, Department of Applied Mathematical Studies, University of Leeds, Leeds, LS2 9JT England

**L. Bos**, Department of Mathematics and Statistics, University of Calgary, Calgary, Alberta, Canada T2N 1N4

**Guido Brunnett**, Department of Computer Science, Arizona State University, Tempe, AZ 85287

**Morton Dæhlen**, Center for Industrial Research, Box 124, Blindern, 0314 Oslo 3, Norway

**Tony DeRose**, Department of Computer Science and Engineering, University of Washington, Seattle, WA 98195

**Matthias Eck**, Fachbereich Mathematik, Technische Hochschule Darmstadt, 6100 Darmstadt, Germany

**Frederick N. Fritsch**, Computing and Math Research Division, Lawrence Livermore National Laboratory, Livermore, CA 94550

**Hans Hagen**, Universität Kaiserslautern, 6750 Kaiserslautern, Germany

**Weimin Han**, Department of Mathematics, University of Iowa, Iowa City, IA 52242

**Emery Jou**, PERFORMIX, Inc., 8521 Leesburg Pike, Vienna, VA 22182

**Dieter Lasser**, Universitat Kaiserslautern, 6750 Kaiserslautern, Germany

**Charles Loop**, Department of Computer Science and Engineering, University of Washington, Seattle, WA 98195

**Michael Lounsbery**, Department of Computer Science and Engineering, University of Washington, Seattle, WA 98195

**Stephen Mann**, Department of Computer Science and Engineering, University of Washington, Seattle, WA 98195

**David Meyers**, Department of Computer Science and Engineering, University of Washington, Seattle, WA 98195

**David A. Neuser**, McDonnell Douglas Helicopter Comp., 5000 E. McDowell Rd., Mesa, AZ 85205

**Gregory M. Nielson**, Department of Computer Science, Arizona State University, Tempe, AZ 85287

**James Painter**, Department of Computer Science, University of Utah, Salt Lake City, UT 84112

**K. Šalkauskas**, Department of Mathematics and Statistics, University of Calgary, Calgary, Alberta, Canada T2N 1N4

**Kenneth Sloan**, Computer and Information Sciences, University of Alabama, Birmingham, AL 35294

**M. J. Wilson**, Department of Applied Mathematical Studies, University of Leeds, Leeds, LS2 9JT England

# Contents

# Curve Design

# Properties of Minimal-Energy Splines

Guido H. Brunnett

## 1.1. Introduction

In the context of computer aided geometric design (CAGD) the energy integral

$$E(x) = \int_0^L \kappa^2(s)\,ds$$

of a plane spline curve $x$ with curvature $\kappa$ and arclength $s$ is often used as a measure of fairness of the curve, although the knowledge about minimal-energy curves is still fragmentary.

In this paper an analysis of the criterion of minimal energy is presented with the intent to establish properties of minimal-energy splines which can be used to evaluate the usefulness of these curves for design purposes.

The layout of this article is as follows. First we recall the differential equation for the curvature function of a minimal-energy curve, which was first derived by Birkhoff and de Boor in 1964 [1] and then verified by Lee and Forsyth in 1974 [7] for the case of segmented curves. The derivation presented emphasizes the use of Erdmann's corner conditions in order to establish the result. It is shown that all solutions of this differential equation are periodic functions which can be represented using the lemniscate function introduced by C.F. Gauss. The relation between the curvature maximum $\kappa_m$ and the period $T$ of a minimal-energy curve is established; an upper bound for the energy functional is also given.

The relation between $\kappa_m$ and $T$ suffices to show the boundedness of the turning angle of a minimal-energy curve. The maximum turning angle is found to be $\pi$ which was already known to Golomb and Jerome [4]. Therefore it is always possible to represent a minimal- energy curve as the graph of a function if the coordinate system is introduced in an appropriate way. This property is used to show the nonexistence of interpolating minimal-energy curves and the appearance of turning points in these curves for certain data configurations.

In §1.5 it is shown that the curvature of a minimal-energy curve $x$ is essentially given by the projection of the vector $x(s) - x(0)$ onto the tangent vector at a turning point of the curve. This projection property is then used to derive the first-order differential equation for the functional case of a minimal-energy curve which was given earlier by Horn [5]. The derivation presented provides an expression for the coefficients of the equation in terms of characteristic curvature values of the curve.

In §1.6 formulas concerning the curvature values $\kappa_0$ and $\kappa_1$ at the end points of the curve are given. A quadratic equation in $\kappa_0$ and $\kappa_1$ is established whose coefficients only depend on the positional and first derivative data at the end points of the curve.

The invariance properties of minimal-energy curves are investigated in §1.7. The property of free elasticity is found to be invariant under uniform scaling and it is shown how to construct scaled interpolants by scaling certain parameters.

In the last section the qualitative behavior of minimal-energy spline is discussed and illustrated by curve plots.

## 1.2. The Differential Equation for a Local Minimum of Energy

First we consider the case of unsegmented curves of class $C^\infty$. Given two points $P$, $Q$ and two unit vectors $V \in T_P\mathbf{R}^2$, $W \in T_Q\mathbf{R}^2$ we define $M$ to be the set of all regular $C^\infty$ curves that interpolate the given data:

$$M := \{x : [a,b] \to \mathbf{R}^2 \;\; : \;\; x \in C^\infty[a,b] \text{ regular}, x(a) = P, x(b) = Q,$$
$$x'(a) = \alpha V, x'(b) = \beta W; \; \alpha, \beta \in \mathbf{R}^+\}.$$

It was first pointed out by Birkhoff and de Boor in 1964 [1] that a curve with minimal energy among all curves of $M$ only exists in the trivial case where a straight line is a possible solution to the problem. This is because we can construct large loops joining given end points with given endslopes, of length $2\pi r$ and curvature $\kappa = O(\frac{1}{r})$, for arbitrarily large $r$. Hence $\int \kappa^2 ds$ can be made smaller than any preassigned positive number. For a more detailed proof see [8].

The absence of an absolute energy minimum can be overcome by a reduction of the set $M$ of comparison curves. If the class of admissible curves is defined as the subset $N_L$ of all curves of $M$ with a prescribed length $L$, then the unique existence of an absolute minimum of energy with respect to $N_L$ can be shown [6]. But fixing the arclength means putting an additional isoperimetric constraint on the variational problem. From the Lagrange multiplier rule it is obvious that the constrained problem

(1.1) $$\int_a^b \kappa^2(t) \, | \, x'(t) \, | \, dt \to \min$$

(1.2) $$\int_a^b | \, x'(t) \, | \, dt = L$$

has the same extremals as the following unconstrained problem

$$(1.3) \qquad \int_a^b (\kappa^2(t) + \lambda^2) \mid x'(t) \mid dt \to \min$$

with some constant $\lambda^2 \in \mathbf{R}^+$. Therefore this approach yields a minimal-energy curve only if the length chosen in (1.2) happens to be the length of an interpolating minimal-energy curve. In all other cases this approach gives a minimal-energy curve in tension [8].

For the investigation of minimal-energy splines without tension we have to introduce the concept of local minima of energy with respect to the set $M$.

DEFINITION 1.2.1. The curve $x \in M$ yields a local minimum of energy with respect to $M$, if there is an open set $O \subset \mathbf{R}^2$ with $x[a, b] \subset O$ and if

$$E(x) \leq E(y)$$

is valid for all $y \in M$ with $y[a, b] \subset O$.

Throughout this paper the term *minimal energy* is used in the sense of a local minimum.

From the viewpoint of mechanics a minimal-energy curve represents a stable interpolating elastica. By the means of calculus of variation a necessary condition for a local minimum can be derived. According to Blaschke [2] a curve which is a local minimizer for the integral

$$R = \int_0^L f(\kappa(s)) ds$$

with respect to $M$ has a curvature function $\kappa$ that satisfies the differential equation

$$(1.4) \qquad \kappa^2 f' + \frac{d^2 f'}{ds^2} - \kappa f = 0.$$

For $f(\kappa) = \kappa^2$, (1.4) yields

$$(1.5) \qquad \kappa''(s) + \frac{1}{2}\kappa^3(s) = 0.$$

This differential equation was first found by Birkhoff and de Boor [1] to be the fundamental equation of an elastica without length constraints. Therefore we introduce the following name.

DEFINITION 1.2.2. An arclength parameterized $C^\infty$ curve with a curvature function that satisfies the differential equation (1.5) is called a *free elastic* curve.

We now consider the more general problem of segmented curves with minimal energy. For the case of spline curves with free end points as well as for closed spline curves, Lee and Forsyth have shown in 1974 [7] that a minimal- energy spline is a curvature-continuous curve with segments that satisfy (1.5). This result remains valid for clamped spline curves (fixed end points and fixed end directions) [8]. We give here a statement of the problem and a short derivation of the results.

Given $n$ points $P_1, P_2, \cdots, P_n \in \mathbf{R}^2$ and two unit vectors $V \in T_{P_1}\mathbf{R}^2, W \in T_{P_n}\mathbf{R}^2$ a curve $x$ is wanted that is a local minimizer of energy in the set $M_n$ of all arclength parameterized curves $y : [0, L] \to \mathbf{R}^2$ ($L$ not fixed) with the properties:

(i) $y \in C^1[0, L]$;

(ii) there is a partition $\triangle_n : 0 = l_1 < l_2 < \cdots < l_n = L$ of the interval $[0,L]$ with $y_i = y \mid_{[l_i, l_{i+1}]} \in C^\infty[l_i, l_{i+1}]$ for $i = 1, \cdots, n - 1$;

(iii) $y(l_i) = P_i$ for $i = 1, \cdots, n$;

(iv) $y'(0) = V, y'(L) = W$.

An admissible curve $y$ can be represented by a segmented function $\Theta = \Theta(s)$ which gives the angle between the tangent vector of y and some fixed direction in the plane. To satisfy (i), we assume $\Theta$ to be of class $C^0[0, L]$ and furthermore

$$\Theta_i = \Theta \mid_{[l_i, l_{i+1}]} \in C^\infty[l_i, l_{i+1}] \text{ for } i = 1, \cdots, n - 1.$$

Hence the relation

$$\Theta_i'(s) = \kappa_i(s) \text{ for } s \in [l_i, l_{i+1}] \text{ and } i = 1, \cdots, n - 1$$

is valid, where $\kappa_i$ denotes the curvature of $y_i$. Now the problem is to minimize the functional

$$I = \sum_{i=1}^{n-1} \int_{l_i}^{l_{i+1}} \left[\Theta_i'(s)\right]^2 ds$$

under the $2(n - 1)$ isoperimetric constraints

$$(1.6) \qquad \int_{l_i}^{l_{i+1}} (\cos \Theta(s), \sin \Theta(s)) \, ds = P_{i+1} - P_i$$

for $i = 1, \cdots, n - 1$, which arise from the interpolation conditions.

If $x \in M_n$ with $x'(s) = (\cos \theta(s), \sin \theta(s))$ is a local minimizer of I, then we consider the subset of curves in $M_n$ which deviate from $x$ only on the $i$th segment. Then it is obvious that $x$ has to be an extremal of the integral

$$I_i = \int_{l_i}^{l_{i+1}} F_i(s, \Theta_i, \Theta_i') ds$$

with

$$F_i = (\Theta_i')^2 + \lambda_i \cos \Theta_i + \mu_i \sin \Theta_i$$

where the Lagrange multipliers $\lambda_i, \mu_i$ are constants because of the isoperimetric nature of the problem. Therefore $x_i$ satisfies the Euler equation

$$(1.7) \qquad -\lambda_i \sin \theta_i + \mu_i \cos \theta_i - 2\theta_i'' = 0$$

from which the equation

$$\kappa_i'' + \frac{1}{2}\kappa_i^3 = C_i\kappa_i$$

can be easily derived (see [4], [6]–[8]).

As $\Theta$ is only required to be of class $C^0[0, L]$, we have to satisfy the corner conditions of Erdmann at the interior points of the partition $\triangle_n$. Using the abbreviations

$$p_i^- := (l_i, \theta_i(l_i), \theta_i'(l_i))$$
$$p_i^+ := (l_{i+1}, \theta_i(l_{i+1}), \theta_i'(l_{i+1})),$$

the first condition of Erdmann

$$\frac{\partial F_i}{\partial \Theta_i'}(p_i^+) = \frac{\partial F_{i+1}}{\partial \Theta_{i+1}'}(p_{i+1}^-) \quad \text{for } i = 1, \cdots, n-2$$

yields the equations

$$\kappa_i(l_{i+1}) = \kappa_{i+1}(l_{i+1}) \quad \text{for } i = 1, \cdots, n-2,$$

that is, the curvature continuity of the curve.

The second corner condition

$$F_i(p_i^+) - \frac{\partial F_i}{\partial \Theta_i'}(p_i^+)\theta_i'(l_{i+1}) = F_{i+1}(p_{i+1}^-) - \frac{\partial F_{i+1}}{\partial \Theta_{i+1}'}(p_{i+1}^-)\theta_{i+1}'(l_{i+1})$$

for $i = 1, \cdots, n-2$ yields the continuity of the second derivatives of the curvature at the interior partition points:

$$\kappa_i''(l_{i+1}) = \kappa_{i+1}''(l_{i+1}) \quad \text{for } i = 1, \cdots, n-2.$$

If $\kappa_i(l_{i+1}) \neq 0$ for $i = 1, \cdots, n-2$, the continuity conditions immediately give

$$C_i = C_{i+1} =: C \quad \text{for } i = 1, \cdots, n-2.$$

If $\kappa_i(l_{i+1}) = 0$ for some $i$, the identity of the constants $C_i$ can be deduced using the continuity of $\kappa''/\kappa$ at $l_{i+1}$ for $i = 1, \cdots, n-2$.

As the length $L$ of the admissible curves is free to vary, we have also to consider the following end condition (see [3]):

$$F_{n-1}(p_{n-1}^+) - \frac{\partial F_{n-1}}{\partial \Theta_{n-1}'}(p_{n-1}^+)\theta_{n-1}'(l_n) = 0.$$

which finally yields

$$C = 0.$$

We summarize these results in the following theorem.

THEOREM 1.2.1. *The curvature function of an arclength parameterized minimal-energy spline with respect to the set $M_n$ is a continuous function that satisfies* (1.5).

In other words, a minimal-energy spline is a curvature-continuous patched curve of free elastic curve segments.

## 1.3.  Periodicity of Curvature Function

The differential equation (1.5) is the equation of a nonlinear oscillator of type

$$(1.8) \qquad\qquad y'' + g(y) = 0$$

with the odd function $g(y) = \frac{1}{2}y^3$. Under certain assumptions, the periodicity of the solutions of (1.8) can be shown (see [9], [8]). This leads us to the following theorem.

THEOREM 1.3.1. *The curvature function $\kappa$ of a free elastic curve $x$ : $[0, L] \to \mathbf{R}^2$ has the following properties:*

*(i) $\kappa$ is concave (convex) in $]a, b[$ , if $\kappa$ is positive (negative) in $]a, b[$. In particular, the turning points of $x$ coincide with the turning points of $\kappa$.*

*(ii) $\kappa$ can be extended to a periodic function $k$ on the whole real line, with the period*

$$T = \frac{2\lambda}{\kappa_m}$$

*where*

$$\lambda := \frac{(\Gamma(1/4))^2}{\sqrt{2\pi}}, \qquad \kappa_m := \max_{s \in \mathbf{R}} | k(s) | ,$$

*and $\Gamma$ denotes the Gamma function.*

*(iii) If $s_n$ is a zero of $\kappa$, then*

$$\kappa(s_n + s) = -\kappa(s_n - s) .$$

*(iv) If $s_m$ is a zero of $\kappa'$, then*

$$\kappa(s_m + s) = \kappa(s_m - s) .$$

*Proof.* (i) follows directly from (1.5).

(ii)–(iv) As $g(y) = \frac{1}{2}y^3$ is of class $C^1(\mathbf{R})$, each solution $\kappa$ of (1.5) can be extended to the whole real line. Now we use the following result from the theory of differential equations (see [9]):

*If $g(y) > 0$ for $y > 0$ and $\underline{\lim}_{y \to \infty} g(y) > 0$, then each solution of (1.8) is a periodic function with the following symmetry properties:*

$$y(c + x) = -y(c - x) \quad for \quad y(c) = 0,$$

$$y(d + x) = y(d - x) \quad for \quad y'(d) = 0.$$

As $g(y) = \frac{1}{2}y^3$ satisfies the assumptions of the lemma above, (iii), (iv), and the first part of (ii) is proven. Because of (i), (iii), (iv) and the differentiability of $\kappa$, it is obvious that the period can be computed as the double length of the largest interval $[\alpha, \beta]$ on which $\kappa'$ is nonnegative.

If $\kappa' \geq 0$ on $[\alpha, \beta]$, the differential equation (1.5) is equivalent to the first-order equation

$$(1.9) \qquad\qquad \kappa' = \frac{1}{2}\sqrt{\kappa_m^4 - \kappa^4}$$

with $\kappa_m = \max_{s\in[a,a+T]} |\,k(s)\,|$. Separation of the variables and integration yields

$$
(1.10) \qquad \int_{y_\alpha}^{y} \frac{d\tilde{y}}{\sqrt{1-\tilde{y}^4}} = \frac{1}{2}\kappa_m(s-\alpha)
$$

with $y = \kappa/\kappa_m$ and $y_\alpha = \kappa(\alpha)/\kappa_m$. Hence (1.10) yields

$$
\begin{aligned}
T = 2(\beta-\alpha) &= \frac{4}{\kappa_m}\lim_{t\to 1}\int_{-t}^{t}\frac{dy}{\sqrt{1-y^4}} \\
&= \frac{2(\Gamma(1/4))^2}{\kappa_m\sqrt{2\pi}}
\end{aligned}
$$

The periodicity of the solutions of (1.5) can be understood in terms of elliptic functions.

THEOREM 1.3.2. *The curvature function* $\kappa : [0,L] \to \mathbf{R}$ *of a free elastic curve in* $\mathbf{R}^2$ *is given by*

$$
(1.11) \qquad \kappa(s) = \kappa_m \,\mathrm{sl}(\kappa_m(s-\alpha)/2), \qquad s \in [0,L].
$$

*Proof.* In case $\kappa' \geq 0$ on $[\alpha,\beta]$, we get from (1.10) using the symmetry of the integrand

$$
\int_{0}^{y}\frac{d\tilde{y}}{\sqrt{1-\tilde{y}^4}} = \frac{1}{2}\kappa_m(s-\alpha) + \underbrace{\int_{0}^{y_\alpha}\frac{d\tilde{y}}{\sqrt{1-\tilde{y}^4}}}_{C_\alpha}.
$$

Applying the definition of the lemniscate function

$$
\mathrm{sl}(\Phi) = x \iff \int_{0}^{x}\frac{dt}{\sqrt{1-t^4}} = \Phi
$$

yields

$$
\kappa(s) = \kappa_m \,\mathrm{sl}(\kappa_m(s-\alpha)/2 + C_\alpha).
$$

In the same way we achieve for $\kappa' \leq 0$ on $[\alpha,\beta]$:

$$
\kappa(s) = \kappa_m \,\mathrm{cl}(\kappa_m(s-\alpha)/2 + K_\alpha).
$$

with

$$
K_\alpha = \int_{y_\alpha}^{1}\frac{d\tilde{y}}{\sqrt{1-\tilde{y}^4}}
$$

and

$$
\mathrm{cl}(\Phi) = x \iff \int_{x}^{1}\frac{dt}{\sqrt{1-t^4}} = \Phi.
$$

According to Theorem 1.3.1 the extension $k$ of $\kappa$ has a zero $k(\alpha) = 0$ with $k(s)' \geq 0$ for $s \in [\alpha - T/4, \alpha + T/4]$. Therefore

$$
\begin{aligned}
k(s) &= \kappa_m \,\mathrm{sl}(\kappa_m(s-\alpha+T/4)/2 + C_{\alpha-T/4}) \\
(1.12) \qquad &= \kappa_m \,\mathrm{sl}(\kappa_m(s-\alpha)/2)
\end{aligned}
$$

because $C_{\alpha-T/4} = -\kappa_m T/8$. For $s \in [\alpha + T/4, \alpha + 3T/4]$ we get

$$
\begin{aligned}
k(s) &= \kappa_m \, \mathrm{cl}(\kappa_m(s - \alpha - T/4)/2 + K_{\alpha+T/4}) \\
&= \kappa_m \, \mathrm{cl}(\kappa_m(s - \alpha)/2 - \kappa_m T/8) \\
&= \kappa_m \, \mathrm{sl}(\kappa_m(s - \alpha)/2)
\end{aligned}
$$

using $K_{\alpha+T/4} = 0$ and the relation between the functions sl and cl (see [10]). Applying the periodicity of these functions, we verify (1.12) for arbitrary $s$ and we obtain (1.11) by restriction of $k$ to the interval $[0, L]$.

The periodicity can be used to establish an upper bound for the energy of a free elastic curve:

COROLLARY 1.3.1. *For the energy $E$ of a free elastic curve $x : [0, L] \to \mathbf{R}^2$ with at most $n$ inflection points in the open interval $]0, L[$, the inequality*

$$
E \leq \frac{(n + 1)^2 \lambda^2}{L}
$$

*holds.*

*Proof.* As $x$ has only $n$ inflection points in $]0, L[$, the length $L$ of $x$ is bounded by

$$
L \leq \frac{n + 1}{2} T \quad \text{with} \quad T = \frac{2\lambda}{\kappa_m} .
$$

From this we get

$$
\kappa_m \leq \frac{(n + 1)\lambda}{L}
$$

and therefore

$$
E = \int_0^L \kappa^2(s) \, ds \leq \kappa_m^2 L \leq \frac{(n + 1)^2 \lambda^2}{L}.
$$

## 1.4.  Graph of a Function

According to §1.3 the curvature function $\kappa$ of a free elastic curve $x$ is the restriction of a periodic function consisting of convex arcs to some interval $[0, L]$. As the turning angle $\Psi$ of the tangent vector of x is given by

$$
\Psi(s) = \int_0^s \kappa(\tilde{s})d\tilde{s},
$$

the largest possible turning angle $\Psi_{\max}$ of $x$ is determined by the area below one complete convex arc of the function $\kappa$.

Obviously a first upper bound for

$$
\Psi_{\max} = \left| \int_a^{a+T/2} \kappa(s) \, ds \right|
$$

with $\kappa(a) = 0$ is

$$
\Psi_{\max} < \kappa_m T/2 = \lambda \approx 300.46°.
$$

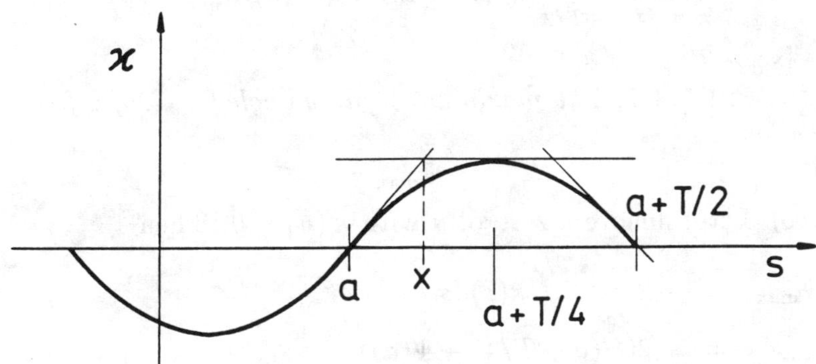


FIG. 1.1. *3-tangent construction.*

Using the 3-tangent construction of Fig. 1.1 we achieve the refined upper bound

$$\Psi_{\max} < \lambda - 2 \approx 185.87°$$

which is close to the exact value claimed in Theorem 1.4.1. To prove this theorem we need the relation between $\kappa$ and $\Psi$ given below.

If $s_m$ denotes an extremum of the extension $k$ of $\kappa$, i.e., if $k'(s_m) = 0$, then we call $\phi = \int_0^{s_m} k(s)ds$ the phase of the free elastic curve with curvature function $\kappa$. Note that $\phi$ is independent from the choice of $s_m$.

LEMMA 1.4.1. *The curvature function $\kappa$ of a free elastic curve with $\kappa_m \neq 0$ and phase $\phi$ satisfies the differential equation*

$$(1.13) \qquad \frac{\kappa^2(s)}{\kappa_m^2} = \cos(\Psi(s) - \phi).$$

*Proof.* Suppose $\kappa' \leq 0$ on $[s_m, s_m + T/2]$, then from $\kappa' = -(1/2)\sqrt{\kappa_m^4 - \kappa^4}$ we get

$$\kappa(s) = -\frac{2\kappa(s)\kappa'(s)}{\kappa_m^2\sqrt{1 - \kappa^4(s)/\kappa_m^4}}.$$

Integration yields for $s \in [s_m, s_m + T/2]$

$$\begin{aligned}
\int_{s_m}^s \kappa(t)dt &= \int_{s_m}^s -\frac{2\kappa(t)\kappa'(t)}{\kappa_m^2\sqrt{1 - \kappa^4(t)/\kappa_m^4}}dt \\
&= \arccos(\kappa^2(s)/\kappa_m^2)
\end{aligned}$$

because $\kappa^2(s_m) = \kappa_m^2$. Hence we have

$$\begin{aligned}
\Psi(s) &= \int_0^{s_m} \kappa(t)dt + \int_{s_m}^s \kappa(t)dt \\
&= \phi + \arccos(\kappa^2(s)/\kappa_m^2).
\end{aligned}$$

For $\kappa'(s) \geq 0$ we proceed in the same way using

$$\int_{s_m+T/2}^{s} \kappa(t)dt = \arccos(\kappa^2(s)/\kappa_m^2) - \pi/2$$

for $s \in [s_m + T/2, s_m + T]$.

THEOREM 1.4.1. *The maximum turning angle* $\Psi_{\max}$ *of a free elastic curve is given by*

(1.14)
$$\Psi_{\max} = \pi.$$

*Proof.* Let $a$ denote a zero of $\kappa$ with $\kappa'(a) \leq 0$. Then

$$\begin{aligned}
\Psi_{\max} &= -2\int_{a}^{a+T/4} \kappa(s)ds \\
&= -2(\Psi(a + T/4) - \Psi(a)) \\
&= -2\left(\phi + \arccos\left(\frac{\kappa^2(a+T/4)}{\kappa_m^2}\right)\right) - \phi - \arccos\left(\frac{\kappa^2(a)}{\kappa_m^2}\right) \\
&= -2(\arccos 1 - \arccos 0) = \pi.
\end{aligned}$$

As an immediate conclusion from (1.14) we obtain

COROLLARY 1.4.1. *A free elastic curve can always be represented as the graph of a function when the coordinate system* $(u, v)$ *is chosen in such a way that the* $v$-*axis is parallel to a tangent in a turning point of the curve.*

Theorem 1.4.1 can also be used to derive qualitative statements about interpolating free elastic curves. For simplicity we assume that the curves $x$ considered interpolate data of the following form:

(1.15)
$$\begin{aligned}
x(0) = 0, \quad x(L) = (P^1, P^2) \quad \text{with } P^1 \geq 0, \\
x'(0) = (0, 1), \quad x'(L) = (-\sin\omega, \cos\omega).
\end{aligned}$$

This choice of the data does not involve any loss of generality because of the invariance properties of free elastic curves which are studied in the next section. The angle $\omega$ gives the oriented angle between $x'(0)$ and $x'(L)$. Let $Q^+$ and $Q^-$ denote the subsets of points $P = (P^1, P^2) \in \mathbf{R}^2$ with $P^1 \geq 0$ and $P^2 \geq 0$ or $P^2 \leq 0$, respectively.

THEOREM 1.4.2. (i) *If* $P^2 \leq 0$ *and* $\pi/2 \leq \omega < \pi$, *then there is no free elastic curve with* (1.15).

(ii) *Necessary for the existence of a free elastic curve with* (1.15) *is the appearance of a turning point in each of the following situations:*

(1.16)
$$0 \leq \omega < \pi$$

(1.17)
$$3\pi/2 \leq \omega < 2\pi \quad and \quad P^2 \leq 0.$$

*Proof.* (i) Let $n_1, n_2, \cdots, n_k$ denote the zeros of $\kappa(s)$ in the open interval $]0, L[$. Then we split the turning angle $\Psi(L)$ into the following three integrals:

$$\Psi(L) = \underbrace{\int_0^{n_1} \kappa(s)\,ds}_{z_1} + \underbrace{\int_{n_1}^{n_k} \kappa(s)\,ds}_{z_2} + \underbrace{\int_{n_k}^{L} \kappa(s)\,ds}_{z_3} =: z\,.$$

Note that $z_1, z_2$ and $z$ lie in $[-\pi, \pi]$, while $z_2 \in \{-\pi, 0, \pi\}$.

Suppose $x$ is a free elastic curve satisfying (1.15) and with negative curvature $\kappa_0$ at $x(0)$. Then this implies that $z_1 \in [-\pi, 0]$ and $z_2 \in \{0, \pi\}$. If $\Psi(n_1) > -\pi/2$ the tangent of $x$ at the turning point $x(n_1)$ separates the region $Q^-$ (which lies "below" the tangent) from the region that contains curve points $x(s)$ with $s > n_1$ (this region lies "above" the tangent). Therefore a curve point $x(s) \in Q^-$ can only exist, if $z_1 = -\pi/2 - \epsilon$ with $\epsilon > 0$.

An arbitrary curve with a turning angle less than $2\pi$ that interpolates the data (1.15) satisfies either $\Psi(L) = \omega$ or $\Psi(L) = -2\pi + \omega$. For a free elastic curve, however, the second case is excluded because $-2\pi + \omega \in \, ] - \pi, -3\pi/2[$ in this situation. Therefore we obtain the equation

$$(1.18) \qquad \omega = \pi/2 + \delta = z = z_1 + z_2 + z_3 \quad \text{with} \quad \delta > 0.$$

Using $z_1 = -\pi/2 - \epsilon$, we get

$$z_2 + z_3 = \pi + \epsilon + \delta,$$

which is a contradiction to the properties of $z_2$ and $z_3$.

Suppose $x$ is a free elastic curve with (1.15) and positive curvature $\kappa_0$ at $x(0)$. Then this implies that $z_1 \in [0, \pi]$ and $z_2 \in \{0, -\pi\}$. Using the same argument as in the first case, we show that $z_1$ has to be of the form $z_1 = \pi/2 - \epsilon$. Equation (1.18) yields in this case:

$$(1.19) \qquad z_2 + z_3 = \epsilon + \delta > 0.$$

If $z_2 = 0$, $z_3$ has to be negative (because of the symmetry properties of $\kappa$), which is a contradiction to (1.19). If $z_2 = -\pi$, (1.19) yields $z_3 > \pi$ which is also a contradiction.

(ii) A curve $x$ with negative curvature throughout in the situation (1.15) has a turning angle $\Psi(L) = -2\pi + \omega$. If $\omega \in [0, \pi[$, $x$ is not a free elastic curve. If $\omega \in [3\pi/2, 2\pi[$, then $x(L) \in Q^-$. A curve with positive curvature throughout and $\omega \in \,]0, \pi[$ has an endpoint $x(L)$ with $x^1(L) < 0$.

## 1.5.  Projection Property

By integrating the Euler equation (1.7) of a general elastica, the relation

$$\kappa(s) = -(\lambda/2)(x^2(s) - x^2(0)) + (\mu/2)(x^1(s) - x^1(0)) + \kappa(0)$$

can be derived (see [7]). In the following we determine the unknown coefficients $\lambda$ and $\mu$ for the case of a free elastic curve and show that the above equation can be interpreted as a projection property.

THEOREM 1.5.1. *Let $x$ be a free elastic curve with curvature $\kappa$ and phase $\phi$. Let $\vartheta$ denote the oriented angle between $x'(0)$ and the positive $u$-axis of the coordinate system. Then*

$$(1.20) \qquad \kappa(s) - \kappa(0) \;=\; -(\kappa_m^2/2)(-\sin(\phi + \vartheta)(x^1(s) - x^1(0)) \\ + \cos(\phi + \vartheta)(x^2(s) - x^2(0))).$$

*Proof.* In the case $\kappa(s) \geq 0$, (1.13) yields

$$
\begin{aligned}
\kappa(s) - \kappa(0) &= \kappa_m \sqrt{\cos(\Psi(s) - \phi)} - \kappa_m \sqrt{\cos(\Psi(0) - \phi)} \\
&= (\kappa_m/2) \int_{\cos(\Psi(0)-\phi)}^{\cos(\Psi(s)-\phi)} \frac{dz}{\sqrt{z}} \\
&= -(\kappa_m/2) \int_{\Psi(0)}^{\Psi(s)} \frac{\sin(r - \phi)}{\sqrt{\cos(r - \phi)}} dr \quad \text{with } z = \cos(r - \phi) \\
&= -(\kappa_m^2/2) \int_{\Psi(0)}^{\Psi(s)} \frac{\sin(r - \phi)}{\Psi'(r)} dr
\end{aligned}
$$

and therefore

$$
(1.21) \qquad \kappa(s) - \kappa(0) = -(\kappa_m^2/2) \int_0^s \sin(\Psi(t) - \phi) dt
$$

with $r = \Psi(t)$. In the case $\kappa(s) \leq 0$, (1.21) is derived analogously.

From (1.21) we get

$$
\begin{aligned}
\kappa(s) - \kappa(0) &= -(\kappa_m^2/2) \int_0^s \sin(\Psi(t) + \vartheta - (\phi + \vartheta)) dt \\
&= -(\kappa_m^2/2)\{\cos(\phi + \vartheta) \int_0^s \sin(\Psi(t) + \vartheta) dt \\
&\qquad - \sin(\phi + \vartheta) \int_0^s \cos(\Psi(t) + \vartheta) dt\}.
\end{aligned}
$$

Using the representation of plane curves by their turning angle $\Psi$ we verify (1.20).

The geometrical meaning of (1.20) is emphasized by the following notation:

COROLLARY 1.5.1. *If $T(t_n^+)$ denotes the unit tangent vector of $x$ at a turning point $t_n^+$ where the sign of $\kappa$ changes from positive to negative, then*

$$
\kappa(s) = -(\kappa_m^2/2) < T(t_n^+), x(s) - x(0) > +\kappa(0)
$$

*i.e., the curvature of a free elastic curve is essentially given by the projection of the difference vector $x(s) - x(0)$ on the tangent vector at a turning point $t_n^+$.*

*Proof.* If $\phi$ denotes the phase of $x$, then

$$
\Psi(t_n^+) = \int_0^{t_n^+} \kappa(s) ds = \pi/2 + \phi.
$$

For the angle $\omega$ between $x'(t_n^+)$ and the positive $u$-axis we get therefore

$$
(1.22) \qquad \omega = \Psi(t_n^+) + \vartheta = \pi/2 + \phi + \vartheta.
$$

Hence we have

$$
(-\sin(\phi + \vartheta), \cos(\phi + \vartheta)) = (\cos\omega, \sin\omega) = T(t_n^+).
$$

For $\vartheta = -\phi$ the coordinate system $(u, v)$ is chosen in such a way that $T(t_n^+)$ is parallel to the positive $v$-axis because from (1.22) we get in this situation: $\omega = \pi/2$. According to Corollary 1.4.1 it is now possible to represent the curve $x$ as the graph of a function $f$. Using the projection property we derive the differential equation of this function.

THEOREM 1.5.2. *Let $x$ be a free elastic curve with curvature $\kappa$ and phase $\phi$. If the oriented angle $\vartheta$ between $x'(0)$ and the positive $u$-axis is given by $\vartheta = -\phi$ then $x$ can be reparameterized in the form $y(u) = (u, f(u))$, where $f$ satisfies*

$$(f'(u))^2 = \frac{1 - (af(u) + b)^4}{(af(u) + b)^4}$$

*with*

$$a = -\kappa_m/2, \qquad b = (\kappa_m/2)x^2(0) + \kappa_0/\kappa_m.$$

*Proof.* In the above situation $x$ is the graph of a function $f$ according to Corollary 1.4.1. The slope of this function at $u$ is the tangent of the angle between $x'(s)$ (with $x(s) = y(u)$) and the positive $u$-axis. Hence we have

$$\begin{aligned}
(f'(u))^2 &= \tan^2(\Psi(s) + \vartheta) \\
&= \frac{1 - \cos^2(\Psi(s) + \vartheta)}{\cos^2(\Psi(s) + \vartheta)}.
\end{aligned}$$

Using the projection property together with (1.13) we achieve

$$\begin{aligned}
\cos^2(\Psi(s) + \vartheta) &= \cos^2(\Psi(s) - \phi) \\
&= [-\kappa_m x^2(s)/2 + \kappa_m x^2(0)/2 + \kappa_0/\kappa_m]^4.
\end{aligned}$$

As $x^2(s) = f(u)$ this gives the claimed result.

## 1.6. The Curvatures at the End Points

As the curvature information at the end points of the curve is extremely useful in the context of curve modeling we derive in this section formulas concerning these curvature values.

LEMMA 1.6.1. *For a free elastic curve with curvature $\kappa$ ($\kappa_0 = \kappa(0)$), turning angle $\Psi$ and phase $\phi$ the following relations hold*

$$(1.23) \qquad\qquad \kappa_m^2 \cos\phi = \kappa_0$$

$$(1.24) \qquad\qquad \kappa_m^2 \sin\phi = \frac{\kappa^2(s) - \kappa_0^2 \cos\Psi(s)}{\sin\Psi(s)} \quad \text{for } \sin\Psi(s) \neq 0$$

$$(1.25) \qquad\qquad \kappa_m^2 \sin\phi = \pm\kappa_m^2 \quad \text{for } \sin\Psi(s) = 0.$$

*Proof.* Equation (1.23) can be deduced directly from (1.13) with $\Psi(0) = 0$. Equation (1.24) follows from (1.13) by applying the addition properties of the trigonometric functions together with (1.23). In the special case $\sin\Psi(s) = 0$ we have $\phi = \pm\pi/2$ according to Theorem 1.4.1 or $\kappa_m = 0$.

The projection property together with the above lemma yield the following important relation:

THEOREM 1.6.1. *For the curvatures $\kappa_0 = \kappa(0), \kappa_1 = \kappa(L)$ of a free elastic curve $x : [0, L] \to \mathbf{R}^2$ the following relation holds:*

$$(1.26) \qquad\qquad C\kappa_1^2 - D\kappa_1 - E\kappa_0^2 + D\kappa_0 = 0$$

*with*

$$C = \cos\vartheta(x^1(L) - x^1(0)) + \sin\vartheta(x^2(L) - x^2(0))$$
$$D = 2\sin\Psi(L)$$
$$E = \cos(\Psi(L) + \vartheta)(x^1(L) - x^1(0)) + \sin(\Psi(L) + \vartheta)(x^2(L) - x^2(0))$$

*where $\vartheta$ denotes the angle between $x'(0)$ and the positive u-axis.*

*Proof.* If $\sin\Psi(s) \neq 0$ then (1.26) can be verified applying the addition properties of the trigonometric functions appearing in the Projection Theorem and using (1.23). If $\sin\Psi(s) = 0$ then (1.26) is trivially satisfied.

The quadratic equation (1.26) obviously allows us to compute the curvature in one endpoint of the curve if the curvature at the other endpoint is known.

It is also possible to express the energy of a free elastic curve as a function of the endpoint curvatures.

THEOREM 1.6.2. *The energy $E$ of a free elastic curve $x : [0, L] \to \mathbf{R}^2$ with curvatures $\kappa_0 = \kappa(0), \kappa_1 = \kappa(L)$ is given by*

$$(1.27) \qquad\qquad E(\kappa_0, \kappa_1) = A\kappa_0^2 + B\kappa_1^2$$

*with*

$$\begin{aligned}
A = \quad & \cos\vartheta(x^1(L) - x^1(0)) + \sin\vartheta(x^1(L) - x^1(0)) \\
- \quad & \cot\Psi(L)\{\cos\vartheta(x^2(L) - x^2(0)) - \sin\vartheta(x^1(L) - x^1(0))\} \\
B = \quad & \{\cos\vartheta(x^2(L) - x^2(0)) - \sin\vartheta(x^2(L) - x^2(0))\}/\sin\Psi(L)
\end{aligned}$$

*for $\sin\Psi(L) \neq 0$. If $\sin\Psi(L) = 0$ then*

$$(1.28) \qquad\qquad E = \kappa_m^2|x(L) - x(0)|.$$

*Proof.* First we get using (1.13),

$$\begin{aligned}
E = \int_0^L \kappa^2(s)ds \;\; = \;\; & \kappa_m^2 \int_0^L \cos(\Psi(s) + \vartheta - (\phi + \vartheta))ds \\
= \;\; & \kappa_m^2\{\cos(\phi + \vartheta)(x^1(L) - x^1(0)) \\
(1.29) \qquad\qquad + \;\; & \sin(\phi + \vartheta)(x^2(L) - x^2(0))\}
\end{aligned}$$

Using the addition properties of the functions in (1.29) together with (1.23) and (1.24) yields (1.27). (1.28) is easily derived from (1.29) using a special coordinate system according to $\vartheta = -\phi$.

If the energy of a free elastic curve is known the curvatures at the end points of the curve can be computed solving a system of two quadratic equations given by (1.26) and (1.27).

## 1.7. Invariance Properties

As the curvature is a geometric quantity of a curve the property of free elasticity
is invariant under translations and rotations. It is easy to show that this
property remains also unchanged if a reflection or a change of orientation is
applied to the curve.

Of major importance for the computation of minimal-energy curves is the
scale invariance property, which can be used to simplify the interpolation
problem:

If $x = x(s_x)$ is a curve that interpolates the data:

$$(1.30) \qquad \begin{aligned} &x\,(0) = 0, x\,(L_x) = P \quad \text{with} \quad |\,P\,| = 1 \\ &x'(0) = V, x'(L_x) = W, \end{aligned}$$

then the scaled curve $y = y(s_y)$ defined by

$$y(ds_x) = dx(s_x)$$

obviously interpolates the scaled position data where the derivatives remain
unchanged:

$$(1.31) \qquad \begin{aligned} &y\,(0) = 0, y(dL_x) = dP \\ &y'(0) = V, y'(dL_x) = W. \end{aligned}$$

The following lemma gives the relation between the curvature functions of $x$
and $y$.

LEMMA 1.7.1. *If $x = x(s_x)$ and $y = y(s_y)$ are two plane curves with
$x(0) = y(0) = 0$ and $x'(0) = y'(0)$ then the following equations are equivalent:*

$$(1.32) \qquad y(ds_x) = dx(s_x)$$
$$(1.33) \qquad \kappa_y(ds_x) = (1/d)\kappa_x(s_x)$$

*Proof.* (a) Differentiating (1.32) twice and using the definition of the
curvature function $\kappa_y(s_y) = \det(y'(s_y), y''(s_y))$ yields (1.33).

(b) Integrating (1.33) yields the relation

$$\Psi_y(ds_x) = \Psi_x(s_x)$$

for the turning angles $\Psi_x$ and $\Psi_y$ of the curves $x$ and $y$. Using the
representation of $x$ and $y$ by their turning angles we get

$$\begin{aligned}
y(ds_x) &= \int_0^{ds_x} (\cos(\Psi_y(t) + \vartheta), \sin(\Psi_y(t) + \vartheta))\, dt \\
&= \int_0^{ds_x} \left( \cos\left(\Psi_x\left(\frac{t}{d}\right) + \vartheta\right), \sin\left(\Psi_x\left(\frac{t}{d}\right) + \vartheta\right) \right) dt \\
&= d\int_0^{s_x} (\cos(\Psi_x(r) + \vartheta), \sin(\Psi_x(r) + \vartheta))\, dr \\
&= dx(s_x)
\end{aligned}$$

with the substitution $r = t/d$.

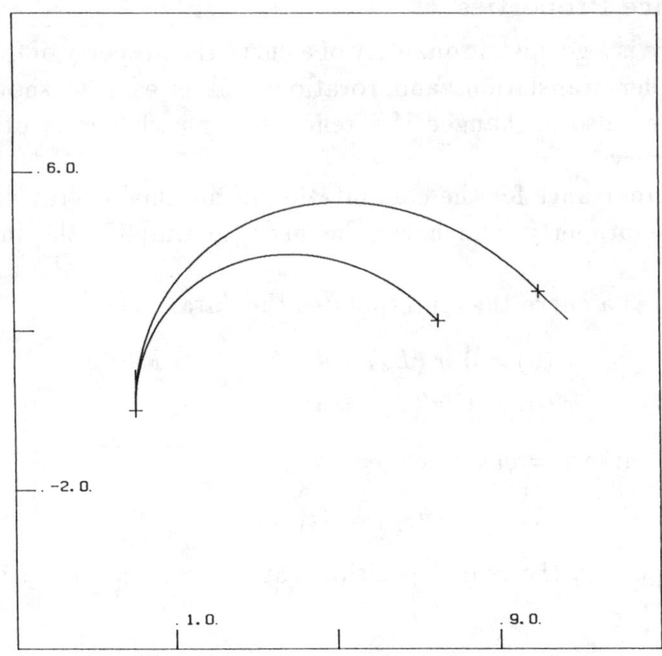


FIG. 1.2. *Scaled free elastic curves.*

The statement of the above lemma can be visualized as follows: if the cuvature $\kappa$ of $x$ with $x(0) = 0$ is plotted versus the arclength $s$ of $x$, then squeezing of $\kappa$ by $\frac{1}{d}$ and stretching of $s$ by $d$ results in a curvature function of a curve $y$ with scaled position data and unchanged turning angle.

The property of free elasticity is invariant under this scaling of the curvature function.

THEOREM 1.7.1. *If* $x = x(s_x)$ *is a free elastic curve with* $x(0) = 0$ *then the scaled curve* $y = y(s_y)$ *with* $y(ds_x) = dx(s_x)$ *is a free elastic curve with energy* $E(y) = (1/d)E(x)$.

*Proof.* According to Lemma 1.7.1 the curvature functions of $x$ and $y$ are related by (1.33). This relation yields the equations

$$\kappa_y''(ds_x) = (1/d)^3 \kappa_x''(s_x)$$

$$\kappa_y^3(ds_x) = (1/d)^3 \kappa_x(s_x).$$

Obviously $\kappa_y$ satisfies the differential equation (1.5) if $\kappa_x$ does. The energy relation is established by substituting $\kappa_y(s_y)$ by $(1/d)\kappa_x(s_y/d)$ at the energy integral $E(y)$.

According to §1.3 the curvature functions of free elastic curves form a two parametric set of periodic functions depending on the parameters $\kappa_m$ and $a$ which specify the amplitude and the phase of the functions. The problem of determining a free elastic curve $x$ that joins two fixed points and takes specified tangent vectors at these points can therefore be reduced to the problem of specifying the parameters $\kappa_m$, $a$ and $L$ where $L$ denotes the arclength of $x$.

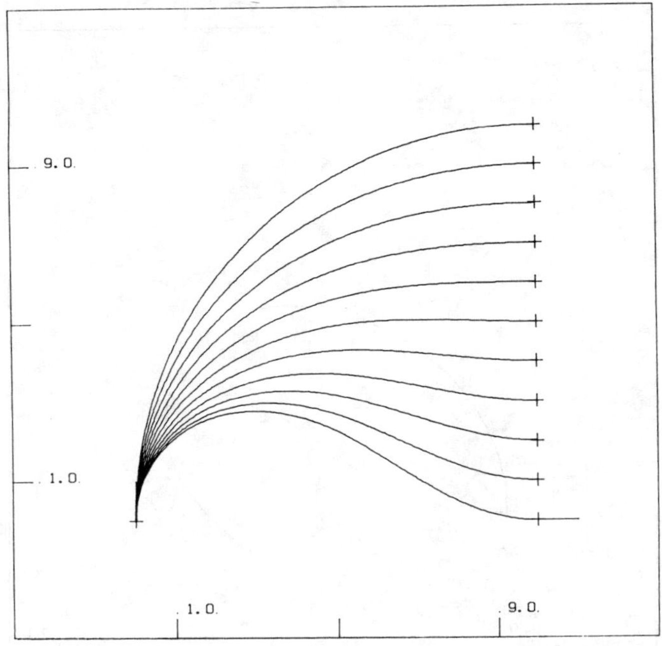



FIG. 1.3.  *Turning points in free elastic curves.*

If the scale invariance property is used to normalize the given data, the parameters of the scaled problem can be obtained by scaling the parameters of the corresponding normalized problem.

COROLLARY 1.7.1. *If $\kappa_m$ and $a$ are the parameters of a free elastic curve $x = x(s_x)$ that interpolates the data (1.30) then a scaled free elastic curve $y = y(s_y)$ with parameters $\tilde{\kappa}_m$ and $\tilde{a}$ that interpolates the data (1.31) can be constructed using the relations:*

$$(1.34) \qquad \tilde{\kappa}_m = \frac{\kappa_m}{d}, \quad \tilde{a} = da, \quad s_y = ds_x.$$

*Proof.* Using (1.34) together with the representations of the curvature functions of $x$ and $y$ according to (1.11) yields the relation (1.33) for the curvature functions of $y$ and $x$. If the initial values of $y$ are specified to be $y(0) = x(0) = 0$ and $y'(0) = x'(0)$, $y$ interpolates the data (1.31) according to Lemma (1.7.1).

## 1.8.  The Shape of Minimal-Energy Splines

The behaviour of minimal-energy curves is guided by the principle of avoiding regions with extreme curvature because large curvature results in a large energy value. This principle creates round-looking shapes (see Figs. 1.4–1.7), which on the one hand please the eye of the observer but on the other hand are unwanted in most technical applications.

To avoid large curvature values it is necessary to introduce a turning point

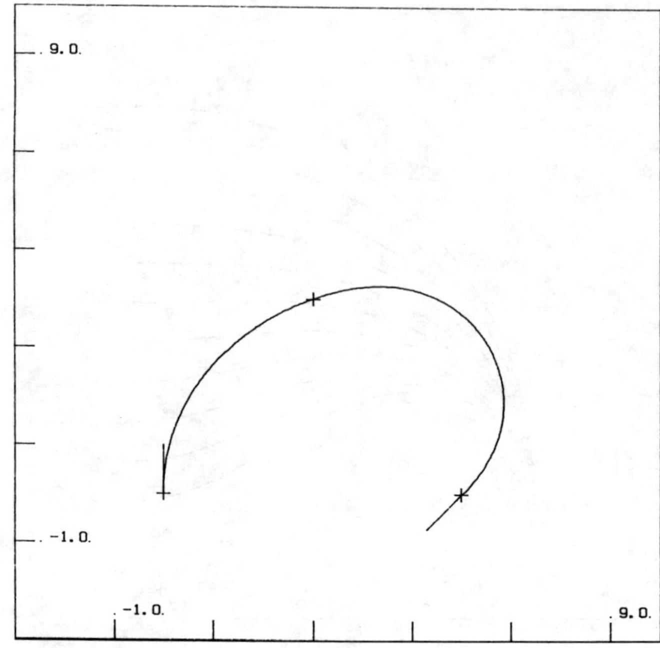

FIG. 1.4.  *Minimal-energy spline in a convex situation.*

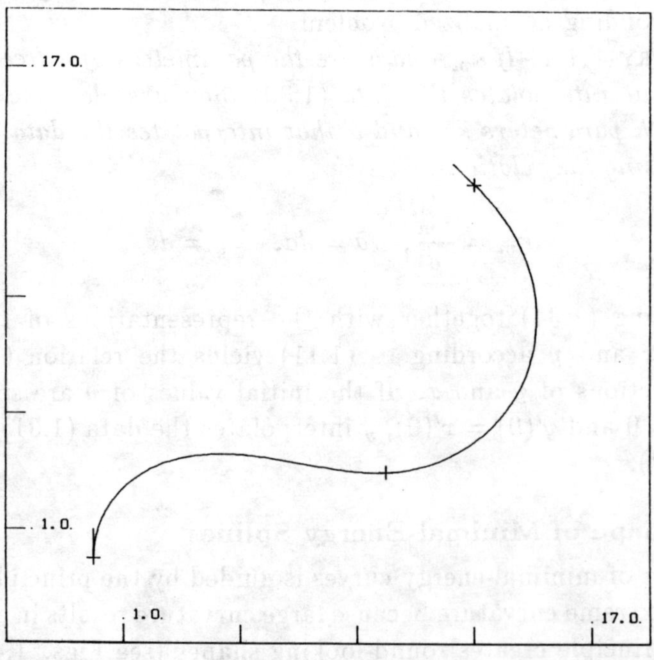

FIG. 1.5.  *Minimal-energy spline in a nonconvex situation.*

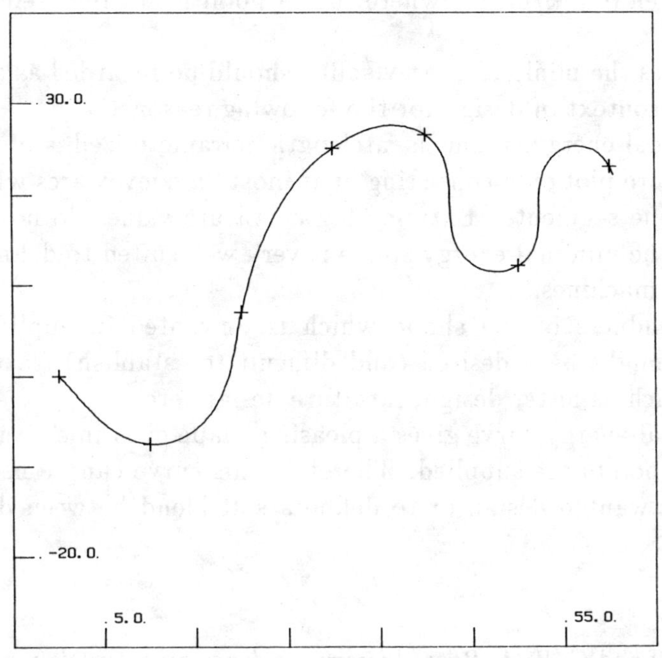

FIG. 1.6.  *Minimal-energy spline: six segments.*

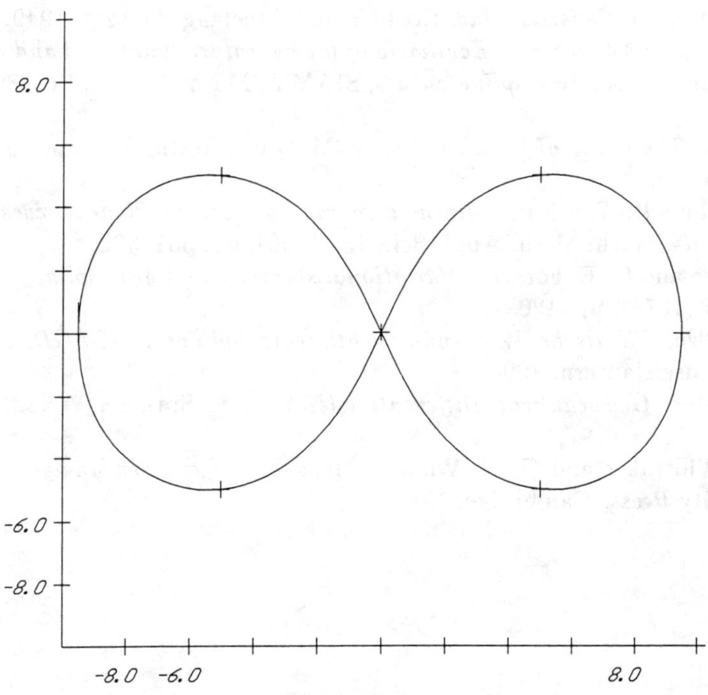

FIG. 1.7.  *Closed minimal-energy spline: eight segments.*

in the curve even in a situation where such a point is not required by the data (see Fig. 1.3).

Nevertheless the minimal-energy spline should be regarded as a useful tool in the general context of design for the following reasons:

• A minimal-energy spline is arclength parameterized and has a very smooth curvature plot only consisting of at most $2n$ convex arcs where $n$ is the number of spline segments. Extreme big curvature values do not occur. For these reasons the minimal-energy spline is very well suited to define curves for NC-controlled machines.

• The roundness of the shape which is unwanted in applications such as car design might be a desired (and difficult to establish) feature in other applications such as letter design, furniture design, etc.

• A minimal-energy curve gives a pleasing shape even in a situation where only few data points are supplied. Therefore this curve can be used to create a preview of a wanted design or to define a soft blend between distinct data sets.

## References

[1] G. Birkhoff and C. R. de Boor, *Piecewise polynomial interpolation and approximation of functions*, General Motors Symposium of 1964, H. L. Garabedian, ed., Elsevier, Amsterdam, 1965, pp. 164–190.

[2] W. Blaschke, *Vorlesungen ueber Differentialgeometrie* I, Chelsea, New York, 1967.

[3] O. Bolza, *Variationsrechnung*, Koehler und Amelang, Leibzig, 1949.

[4] M. Golomb and J. Jerome, *Equilibria of the curvature functional and manifolds of nonlinear interpolating spline curves*, SIAM J. Math. Anal., 13 (1982), pp. 421–458.

[5] B. Horn, *The curve of least energy*, ACM Trans. Math. Software, 9 (1983), pp. 441–460.

[6] E. Jou and H. Wei-Min, *Minimal energy splines:* I. *Plane curves with angle constraints*, Math. Meth. Appl. Sci., 13 (1990), pp. 351–372.

[7] E. H. Lee and G. E. Forsyth, *Variational study of nonlinear spline curves*, SIAM Rev., 15 (1975), pp. 120–133.

[8] G. Schulze, *Elastische Wege und nichtlineare Splines im CAGD*, Dissertation, Univ. Kaiserslautern, 1990.

[9] W. Walter, *Gewoehnliche Differentialgleichungen*, Springer-Verlag, New York, 1976.

[10] E. T. Whittaker and G. N. Watson, *A course of modern analysis*, Cambridge University Press, Cambridge, 1944.

# Minimal-Energy Splines with Various End Constraints

Emery Jou and Weimin Han

## 2.1. Introduction

Minimal-energy splines have been considered by others under different names. Many theoretical and numerical results can be found in the literature [1–8] and [13–18]. We have studied a family of minimal-energy splines in [9] and [10]. New features of this spline family include (1) the length of a minimal-energy spline is fixed; (2) the boundary conditions are of more general types. The results and the methods to obtain them are systematic and general. In [12], point forces are introduced to control the variation of curvature at certain points of a minimal-energy spline.

The aim of this paper is to clearly classify a branch of minimal-energy splines with various end constraints which minimize the integral of the curvature squared among all the curves of a prescribed length, passing through a set of points. Clamped, mixed, or natural minimal-energy splines are obtained when we prescribe both end angles, only one end angle, or no end angles. We present various properties of minimal-energy spline segments in §2.2 and those of minimal-energy splines in §2.3. In §2.4, we consider computational aspects of minimal-energy splines and provide graphs for several minimal-energy splines.

## 2.2. Minimal-Energy Spline Segments

We use arclength parameterized forms for curves to study minimal-energy splines. We denote $f'$ the derivative of a function $f(s)$ with respect to the arclength parameter $s$.

Given two points $P_1 = (x_1, y_1)$, $P_2 = (x_2, y_2)$ on the plane, and a positive number $l$, a minimal-energy spline segment considered in this section is a curve of length $l$, subject to certain constraints at the end points $P_1$ and $P_2$, and which minimizes the energy:

$$(2.1) \qquad E(\kappa) = \int_0^l \frac{1}{2}\kappa(s)^2 ds$$

where $\kappa(s)$ is the curvature of the curve.

On an arclength parameterized curve $\{(x(s), y(s)) \mid 0 \leq s \leq l\}$,

$$(2.2) \qquad\qquad\qquad x'(s)^2 + y'(s)^2 = 1$$

and there is a function $\theta(s)$ such that on the curve:

$$(2.3) \qquad\qquad\qquad x'(s) \;\; = \;\; \cos\theta(s),$$
$$(2.4) \qquad\qquad\qquad y'(s) \;\; = \;\; \sin\theta(s).$$

Obviously, $\theta(s)$ may be interpreted as the tangent direction angle of the curve at the point $(x(s), y(s))$. Note that the value of $\theta(s)$ can be any real number. After introducing the function $\theta(s)$, $0 \leq s \leq l$, the curvature of the curve may be computed through

$$(2.5) \qquad\qquad\qquad \kappa(s)^2 = x''(s)^2 + y''(s)^2 = \theta'(s)^2$$

and coordinates of points on the curve are obtained through

$$(2.6) \qquad\qquad\qquad x(s) \;\; = \;\; x_1 + \int_0^s \cos\theta(s)ds,$$

$$(2.7) \qquad\qquad\qquad y(s) \;\; = \;\; y_1 + \int_0^s \sin\theta(s)ds.$$

Hence, the energy function (2.1) can be rewritten as:

$$(2.8) \qquad\qquad\qquad E(\theta) = \int_0^l \frac{1}{2}\theta'(s)^2 ds$$

and the requirement that the curve connects the two end points $P_1$ and $P_2$ is equivalent to

$$(2.9) \qquad\qquad\qquad \int_0^l \cos\theta(s)ds \;\; = \;\; x_2 - x_1,$$

$$(2.10) \qquad\qquad\qquad \int_0^l \sin\theta(s)ds \;\; = \;\; y_2 - y_1.$$

As for constraints at $P_1$ and $P_2$, three cases may occur.

*Case* 1. The angles at both end points $P_1$ and $P_2$ are prescribed:

$$\theta(0) = \alpha_1, \quad \theta(l) = \alpha_2,$$

where $\alpha_1$, $\alpha_2$ are given numbers.

For this case, we introduce the constraint set:

$$\begin{aligned}
H \;\; &= \;\; H_c(P_1, P_2; \alpha_1, \alpha_2; l) \\
&= \;\; \{\theta \in H^1(0, l) \mid \theta(0) = \alpha_1, \; \theta(l) = \alpha_2, \\
&\qquad \int_0^l \cos\theta(s)ds = x_2 - x_1, \; \int_0^l \sin\theta(s)ds = y_2 - y_1\}.
\end{aligned}$$

*Case* 2. No angles are prescribed at the end points.

The constraint set is

$$
\begin{aligned}
H &= H_n(P_1, P_2; l) \\
&= \{\theta \in H^1(0, l) \mid \theta(0) \in [0, 2\pi], \\
&\quad \int_0^l \cos\theta(s)ds = x_2 - x_1, \ \int_0^l \sin\theta(s)ds = y_2 - y_1\}.
\end{aligned}
$$

Note that we restrict $\theta(0)$ to be in $[0, 2\pi]$. Once we have minimal-energy spline segments with this constraint set, we obtain other minimal-energy spline segments without end point constraints by translating the function $\theta(s)$ with amount $2k\pi$ for integers $k$.

*Case* 3. Angle at only one of the two end points is prescribed.

The constraint set for this case is either

$$
\begin{aligned}
H &= H_{m,1}(P_1, P_2; \alpha_1; l) \\
&= \{\theta \in H^1(0, l) \mid \theta(0) = \alpha_1, \\
&\quad \int_0^l \cos\theta(s)ds = x_2 - x_1, \ \int_0^l \sin\theta(s)ds = y_2 - y_1\};
\end{aligned}
$$

or

$$
\begin{aligned}
H &= H_{m,2}(P_1, P_2; \alpha_2; l) \\
&= \{\theta \in H^1(0, l) \mid \theta(l) = \alpha_2, \\
&\quad \int_0^l \cos\theta(s)ds = x_2 - x_1, \ \int_0^l \sin\theta(s)ds = y_2 - y_1\};
\end{aligned}
$$

where either $\alpha_1$ or $\alpha_2$ is a given real number.

Now we form the following definition.

DEFINITION 2.2.1. A plane curve $\{(x^*(s), y^*(s)) \mid 0 \le s \le l\}$ is said to be a minimal-energy spline segment with constraint set $H$, if

$$
\begin{aligned}
x^*(s) &= x_1 + \int_0^s \cos\theta^*(s)ds, \\
y^*(s) &= y_1 + \int_0^s \sin\theta^*(s)ds,
\end{aligned}
$$

where $\theta^* \in H$ minimizes the energy function among all the admissible elements in $H$:

$$
(2.11) \qquad\qquad E(\theta^*) = \inf\{E(\theta) \mid \theta \in H\}
$$

with the energy:

$$
E(\theta) = \int_0^l \frac{1}{2}\theta'(s)^2 ds.
$$

In particular, when $H = H_c(P_1, P_2; \alpha_1, \alpha_2; l)$, the spline segment is called a clamped minimal-energy spline segment; when $H = H_n(P_1, P_2; l)$, the spline segment is called a natural minimal-energy spline segment; when $H = H_{m,1}(P_1, P_2; \alpha_1; l)$, or $H = H_{m,2}(P_1, P_2; \alpha_2; l)$, the spline segment is called a mixed minimal-energy spline segment.

REMARK 2.2.1. When $H = H_n(P_1, P_2; l)$, we do not have end point constraint. From Theorem 2.2.4 below, we will see that for a natural minimal-energy spline segment, $\theta^{*\prime}(0) = \theta^{*\prime}(l) = 0$. Similarly, for a mixed minimal-energy spline segment with $H = H_{m,1}(P_1, P_2; \alpha_1; l)$, there holds $\theta^{*\prime}(l) = 0$; and with $H = H_{m,2}(P_1, P_2; \alpha_2; l)$, $\theta^{*\prime}(0) = 0$.

First, we have to ensure that the constraint sets are nonempty.

LEMMA 2.2.1. *If $l > \sqrt{(x_2 - x_1)^2 + (y_2 - y_1)^2}$, the distance between $P_1$ and $P_2$, then each of the constraint sets $H_c(P_1, P_2; \alpha_1, \alpha_2; l)$, $H_n(P_1, P_2; l)$, $H_{m,1}(P_1, P_2; \alpha_1; l)$, and $H_{m,2}(P_1, P_2; \alpha_2; l)$ is nonempty, for any $\alpha_1, \alpha_2 \in R$.*

The meaning of Lemma 2.2.1 is clear, it says: as long as the length of the curve is greater than the distance between two end points $P_1$ and $P_2$, it is always possible to find a curve of the given length, connecting the end points, and its end angles can be adjusted to satisfy the end constraints (if there exist end constraints). The proof of the Lemma 2.2.1 can be found in [9] and [10].

Now we can state the main results of this section, concerning properties of minimal-energy spline segments.

THEOREM 2.2.1. *If $l > \sqrt{(x_2 - x_1)^2 + (y_2 - y_1)^2}$, then the constrained minimization problem (2.11) has a solution $\theta^* \in H$.*

In general, a solution $\theta^*$ of (2.11) is not unique, hence a minimal-energy spline segment is not unique. For example, let $P_1 = (0,0)$, $P_2 = (1,0)$, $\alpha_1 = \alpha_2 = 0$, $l > 1$. If $\theta^*(s)$ is a solution of (2.11), then the function $-\theta^*(s)$ is a solution of (2.11) as well and $\theta^* \neq -\theta^*$. What kind of assumption on the input data will make a solution unique is an open question.

THEOREM 2.2.2. *If $\theta^* \in H$ is a solution of the constrained problem (2.11), then there exist $\lambda_1, \lambda_2 \in R$, such that*
*If $H = H_c(P_1, P_2; \alpha_1, \alpha_2; l)$, then*

$$\int_0^l \{\theta^{*\prime}\psi^\prime - (\lambda_1 \sin \theta^* + \lambda_2 \cos \theta^*)\psi\} = 0, \qquad \forall \psi \in H_0^1(0, l).$$

*If $H = H_n(P_1, P_2; l)$, then*

$$\int_0^l \{\theta^{*\prime}\psi^\prime - (\lambda_1 \sin \theta^* + \lambda_2 \cos \theta^*)\psi\} = 0, \qquad \forall \psi \in H^1(0, l).$$

*If $H = H_{m,1}(P_1, P_2; \alpha_1; l)$, then*

$$\int_0^l \{\theta^{*\prime}\psi^\prime - (\lambda_1 \sin \theta^* + \lambda_2 \cos \theta^*)\psi\} = 0, \qquad \forall \psi \in H_{(0}^1(0, l).$$

*If $H = H_{m,2}(P_1, P_2; \alpha_2; l)$ then*

$$\int_0^l \{\theta^{*\prime}\psi^\prime - (\lambda_1 \sin \theta^* + \lambda_2 \cos \theta^*)\psi\} = 0, \qquad \forall \psi \in H_{0)}^1(0, l).$$

In Theorem 2.2.2, we used the following notation for Sobolev spaces:

$$H_{(0}^1(0, l) = \{\theta \in H^1(0, l) \mid \theta(0) = 0\},$$

$$H^1_{0)}(0,l) = \{\theta \in H^1(0,l) \mid \theta(l) = 0\}.$$

THEOREM 2.2.3. *A solution $\theta^* \in H$ of the problem (2.11) is infinitely smooth: $\theta^* \in C^\infty([0,l])$.*

For the case of clamped minimal-energy spline segments, the proof of Theorems 2.2.1–2.2.3 can be found in [9]. For the remaining cases, these Theorems are deduced from more general results in [10]. Note that, since the constraint sets are not linear, the conventional way (as used by several authors) to derive the Lagrange multiplier rule is mathematically incomplete.

From Theorems 2.2.2 and 2.2.3, we conclude that

THEOREM 2.2.4. *Let $\theta^* \in H$ be a solution of the constrained problem* (2.11).

*If $H = H_c(P_1, P_2; \alpha_1, \alpha_2; l)$, then $\theta^*$ is a classical solution of*

$$\theta^{*\prime\prime} + \lambda_1 \sin\theta^* + \lambda_2 \cos\theta^* = 0, \ in \ (0,l),$$
$$\theta^*(0) = \alpha_1, \ \theta^*(l) = \alpha_2,$$
$$\int_0^l \cos\theta^*(s)ds = x_2 - x_1, \ \int_0^l \sin\theta^*(s)ds = y_2 - y_1.$$

*If $H = H_n(P_1, P_2; l)$, then $\theta^*$ is a classical solution of*

$$\theta^{*\prime\prime} + \lambda_1 \sin\theta^* + \lambda_2 \cos\theta^* = 0, \ in \ (0,l),$$
$$\theta^{*\prime}(0) = 0, \ \theta^{*\prime}(l) = 0,$$
$$\int_0^l \cos\theta^*(s)ds = x_2 - x_1, \ \int_0^l \sin\theta^*(s)ds = y_2 - y_1.$$

*If $H = H_{m,1}(P_1, P_2; \alpha_1; l)$, then $\theta^*$ is a classical solution of*

$$\theta^{*\prime\prime} + \lambda_1 \sin\theta^* + \lambda_2 \cos\theta^* = 0, \ in \ (0,l),$$
$$\theta^*(0) = \alpha_1, \ \theta^{*\prime}(l) = 0,$$
$$\int_0^l \cos\theta^*(s)ds = x_2 - x_1, \ \int_0^l \sin\theta^*(s)ds = y_2 - y_1.$$

*If $H = H_{m,2}(P_1, P_2; \alpha_2; l)$, then $\theta^*$ is a classical solution of*

$$\theta^{*\prime\prime} + \lambda_1 \sin\theta^* + \lambda_2 \cos\theta^* = 0, \ in \ (0,l),$$
$$\theta^{*\prime}(0) = 0, \ \theta^*(l) = \alpha_2,$$
$$\int_0^l \cos\theta^*(s)ds = x_2 - x_1, \ \int_0^l \sin\theta^*(s)ds = y_2 - y_1.$$

From the equation

$$(2.12) \qquad \theta^{*\prime\prime} + \lambda_1 \sin\theta^* + \lambda_2 \cos\theta^* = 0, \ in \ (0,l)$$

we may derive various well-known relations.

Integrating (2.12), we obtain the linear curvature relation:

$$\kappa = -\lambda_1 y - \lambda_2 x - C.$$

for a minimal-energy spline segment, where $C$ is a constant.

We differentiate (2.12) to get

(2.13) $$\theta^{*'''} + \lambda_1 \cos \theta^* \cdot \theta^{*'} - \lambda_2 \sin \theta^* \cdot \theta^{*'} = 0.$$

Multiplying (2.12) by $\theta^{*'}$ and integrating, we have for some constant $K$:

(2.14) $$\frac{1}{2}\theta^{*'2} - \lambda_1 \cos \theta^* + \lambda_2 \sin \theta^* = K.$$

Eliminating $\lambda_1$ and $\lambda_2$ from (2.13) and (2.14), we obtain

$$\theta^{*'''} + \frac{1}{2}\theta^{*'3} = K\theta^{*'},$$

i.e.,

$$\kappa'' + \frac{1}{2}\kappa^3 = K\kappa.$$

For a natural minimal-energy spline segment, integrating (2.12) from $s = 0$ to $s = l$, we obtain a relation for the two Lagrange multipliers:

$$\lambda_1(y_2 - y_1) + \lambda_2(x_2 - x_1) = 0.$$

## 2.3.  Minimal-Energy Splines

We denote $\mathbf{P} = (P_0, P_1, \cdots, P_N)$, and $\mathbf{l} = (l_1, \cdots, l_N)$, where $P_i = (x_i, y_i), i = 0, \cdots, N$, and $l_j > 0, j = 1, \cdots, N$, are given.  Furthermore, we denote $L_0 = 0, L_i = \sum_{j=1}^{i} l_j, i = 1, \cdots, N$.  In the following definition, we introduce minimal-energy splines satisfying certain end point constraints, connecting $P_0, P_1, \cdots, P_N$, being of length $l_i$ between $P_{i-1}$ and $P_i$ ($i = 1, \cdots, N$), and minimizing the energy function

$$\int_0^l \frac{1}{2}\kappa(s)^2 ds$$

where $\kappa(s)$ is the curvature of the curve, $l = \sum_{i=1}^{N} l_i$ is the length.

DEFINITION 2.3.1. A plane curve $\{(x^*(s), y^*(s)) \mid 0 \leq s \leq l\}$ is said to be a minimal-energy spline with constraint sets $H$, if

$$x^*(s) = x_0 + \int_0^s \cos \theta^*(s) ds,$$

$$y^*(s) = y_0 + \int_0^s \sin \theta^*(s) ds,$$

where $\theta^* \in H$ minimizes the energy function among all the admissible elements in $H$:

(2.15) $$E(\theta^*) = \inf \{E(\theta) \mid \theta \in H\}.$$

with the energy

$$E(\theta) = \int_0^l \frac{1}{2}\theta'(s)^2 ds.$$

In particular,

(1) The spline is called a clamped minimal-energy spline, if

$$
\begin{aligned}
H &= H_c(\mathbf{P}; \alpha_0, \alpha_N; l) \\
&= \{\theta \in H^1(0,l) \mid \theta^*(0) = \alpha_0, \ \theta^*(l) = \alpha_N, \\
&\qquad \int_{L_{i-1}}^{L_i} \cos\theta(s)ds = x_i - x_{i-1}, \\
&\qquad \int_{L_{i-1}}^{L_i} \sin\theta(s)ds = y_i - y_{i-1}, \ i = 1, \cdots, N\}.
\end{aligned}
$$

(2) The spline is called a natural minimal-energy spline, if

$$
\begin{aligned}
H &= H_n(\mathbf{P}; l) \\
&= \{\theta \in H^1(0,l) \mid \\
&\qquad \int_{L_{i-1}}^{L_i} \cos\theta(s)ds = x_i - x_{i-1}, \\
&\qquad \int_{L_{i-1}}^{L_i} \sin\theta(s)ds = y_i - y_{i-1}, \ i = 1, \cdots, N\}.
\end{aligned}
$$

(3) The spline is called a mixed minimal-energy spline, if

$$
\begin{aligned}
H &= H_{m,1}(\mathbf{P}; \alpha_0; l) \\
&= \{\theta \in H^1(0,l) \mid \theta^*(0) = \alpha_0, \\
&\qquad \int_{L_{i-1}}^{L_i} \cos\theta(s)ds = x_i - x_{i-1}, \\
&\qquad \int_{L_{i-1}}^{L_i} \sin\theta(s)ds = y_i - y_{i-1}, \ i = 1, \cdots, N\}.
\end{aligned}
$$

or

$$
\begin{aligned}
H &= H_{m,2}(\mathbf{P}; \alpha_N; l) \\
&= \{\theta \in H^1(0,l) \mid \theta^*(l) = \alpha_N, \\
&\qquad \int_{L_{i-1}}^{L_i} \cos\theta(s)ds = x_i - x_{i-1}, \\
&\qquad \int_{L_{i-1}}^{L_i} \sin\theta(s)ds = y_i - y_{i-1}, \ i = 1, \cdots, N\}.
\end{aligned}
$$

REMARK 2.3.1. In the definitions of various constraint set, the interpolating conditions:

$$\int_{L_{i-1}}^{L_i} \cos\theta(s)ds = x_i - x_{i-1}, \ \int_{L_{i-1}}^{L_i} \sin\theta(s)ds = y_i - y_{i-1}, \ i = 1, \cdots, N,$$

reflect the fact that the spline is of length $L_i - L_{i-1} = l_i$ between $P_{i-1}$ and $P_i$, $i = 1, \cdots, N$.

We have the following similar results for minimal-energy splines.

THEOREM 2.3.1. *Assume*

$$l_i^2 \; > \; (x_i - x_{i-1})^2 + (y_i - y_{i-1})^2, \qquad i = 1, \cdots, N,$$

*then*

(1) *Each of the constraint sets is nonempty, where $\alpha_0, \alpha_N \in R$ are given.*

(2) *For each of the constraint sets, the constrained minimization problem (2.15) has a solution $\theta^* \in H$.*

(3) *$\theta^*$ is continuously differentiable on $[0, l]$ and is infinitely differentiable on each subintervals $[L_{i-1}, L_i]$, $i = 1, \cdots, N$, i.e.,*

$$(2.16) \qquad \theta^* \in C^1([0, l]) \cap \cap_{i=1}^N C^\infty([L_{i-1}, L_i]).$$

(4) *There exist $2N$ real numbers $\lambda_{i,1}, \lambda_{i,2}$, $i = 1, \cdots, N$, such that*

$$(2.17) \qquad \int_0^l \theta^{*\prime} \cdot \psi' - \sum_{i=1}^N \int_{L_{i-1}}^{L_i} (\lambda_{i,1} \sin \theta^* + \lambda_{i,2} \cos \theta^*) \cdot \psi = 0, \, \forall \psi \in \Psi$$

*where the test function space $\Psi$ is defined as follows: If $H = H_c(\mathbf{P}; \alpha_0, \alpha_N; \mathbf{l})$, then $\Psi = H_0^1(0, l)$. If $H = H_n(\mathbf{P}; \mathbf{l})$, then $\Psi = H^1(0, l)$. If $H = H_{m,1}(\mathbf{P}; \alpha_0; \mathbf{l})$, then $\Psi = H_{(0}^1(0, l)$. If $H = H_{m,2}(\mathbf{P}; \alpha_N; \mathbf{l})$, then $\Psi = H_{0)}^1(0, l)$.*

REMARK 2.3.2. From (2.16) and (2.17), we observe that for a clamped minimal-energy spline, $\theta^*$ is a classical solution of the problem:

$$(2.18) \qquad \theta^{*\prime\prime} + \lambda_{i,1} \sin \theta^* + \lambda_{i,2} \cos \theta^* = 0 \;\; \text{in } (L_{i-1}, L_i), \, i = 1, \cdots, N,$$

$$(2.19) \qquad \theta^*(L_i+) = \theta^*(L_i-), \, \theta^{*\prime}(L_i+) = \theta^{*\prime}(L_i-), \, i = 1, \cdots, N-1,$$

$$(2.20) \qquad \theta^*(0) = \alpha_0, \, \theta^*(l) = \alpha_N,$$

$$(2.21) \qquad \int_{L_{i-1}}^{L_i} \cos \theta(s) ds = x_i - x_{i-1},$$

$$\int_{L_{i-1}}^{L_i} \sin \theta(s) ds = y_i - y_{i-1}, \, i = 1, \cdots, N.$$

We note that (2.19) is the transition condition at $P_i$, $i = 1, \cdots, N-1$, which represents the fact that $\theta^* \in C^1([0, l])$.

For a natural minimal-energy spline, we still have relations (2.18), (2.19), (2.21). However, (2.20) is replaced by

$$\theta^{*\prime}(0) = 0, \quad \theta^{*\prime}(l) = 0.$$

Similarly, when $H = H_{m,1}(\mathbf{P}; \alpha_0; \mathbf{l})$, or $H = H_{m,2}(\mathbf{P}; \alpha_N; \mathbf{l})$, relations (2.18), (2.19), (2.21) hold, whereas (2.20) is replaced by

$$\theta^*(0) = \alpha_0, \quad \theta^{*\prime}(l) = 0$$

or

$$\theta^{*\prime}(0) = 0, \quad \theta^*(l) = \alpha_N,$$

respectively.

The following theorem describes one relation between minimal-energy splines and minimal-energy spline segments, the proof of which is easy and is omitted.

THEOREM 2.3.2. *Each segment* $\{(x^*(s+L_{i-1}), y^*(s+L_{i-1})) \mid 0 \le s \le l_i\}$, $i = 1, \cdots, N - 1$, *of a minimal-energy spline is a clamped minimal-energy spline segment with the constraint set* $H_c(P_{i-1}, P_i; \theta^*(L_{i-1}), \theta^*(L_i); l_i)$, $i = 1, \cdots, N - 1$.

## 2.4.  Computation of Minimal-Energy Splines

We state a simple algorithm to solve the constrained minimization problem (2.11):

$$E(\theta^*) = \inf\{E(\theta) \mid \theta \in H\}.$$

We use the clamped minimal-energy spline segment as an example, i.e., the angles at both end points $P_1$ and $P_2$ are prescribed. For this case, the constraint set:

$$\begin{aligned} H &= H_c(P_1, P_2; \alpha_1, \alpha_2; l) \\ &= \{\theta \in H^1(0, l) \mid \theta(0) = \alpha_1, \theta(l) = \alpha_2, \\ &\quad \int_0^l \cos\theta(s)ds = x_2 - x_1, \int_0^l \sin\theta(s)ds = y_2 - y_1\}. \end{aligned}$$

Let $N$ be a positive integer. Denote $h = l/N$, $s_i = ih$, $0 \le i \le N$; $I_i = [s_{i-1}, s_i]$, $1 \le i \le N$; and the discrete constraint set:

$$\begin{aligned} \bar{H} &= H_c^h(P_1, P_2; \alpha_1, \alpha_2; l) \\ &= \{\theta_h \in H^1(0, l) \mid \theta_h(0) = \alpha_1, \theta_h(l) = \alpha_2, \\ &\quad \theta_h \text{ is a polynomial of degree 1 on } I_i, 1 \le i \le N, \\ &\quad \sum_{i=1}^{N} \frac{h}{2}(\cos\theta_h(s_{i-1}) + \cos\theta_h(s_i)) = x_2 - x_1, \\ &\quad \sum_{i=1}^{N} \frac{h}{2}(\sin\theta_h(s_{i-1}) + \sin\theta_h(s_i)) = y_2 - y_1\}. \end{aligned}$$

Then we approximate the problem (2.11) by

$$\text{Find } \theta_h^* \in H_c^h(P_1, P_2; \alpha_1, \alpha_2; l) \text{ such that}$$
$$E(\theta_h^*) = \inf\{E(\theta_h) \mid \theta_h \in H_c^h(P_1, P_2; \alpha_1, \alpha_2; l)\}.$$

Under the assumption of Theorem 2.2.1, we have the following statements concerning the approximate problem.

(a) There exists $h_0 > 0$, such that for all $h \in (0, h_0]$, $H_c^h(P_1, P_2; \alpha_1, \alpha_2; l) \neq \emptyset$, and the approximate problem has a solution $\theta_h^*$.

(b) Each sequence $\{\theta_h^*\}$ contains a subsequence which converges to $\theta^*$ in $H^1(0, l)$ and in $C^0([0, l])$, for a solution $\theta^*$ of the problem (2.11).

The algorithm and convergence result for clamped minimal-energy spline segment can be modified for both the natural and mixed minimal-energy spline segments.

For the constrained minimization problem (2.15), we have similar algorithms and same convergence results to compute minimal-energy splines with various end constraints. The proof of these results can be found in [11].

Numerical results for both open and closed minimal-energy spline segments and minimal-energy splines follow.

We compute a family of fixed-length *closed* clamped minimal-energy spline segments with the constraint sets $H_c(P_1, P_2; \alpha_1, \alpha_2; l)$, where $P_1 = P_2 = (0, 0)$, $\alpha_1 = 0$, $\alpha_2 \in [\pi, 2\pi]$, and $l = \pi$. Twelve closed minimal-energy spline segments are shown in Fig. 2.1, with their corresponding energies listed in Table 2.1. When $\alpha_2 = 2\pi$, the corresponding spline segment is a circle of radius 0.5. The light solid curve corresponds to the case $\alpha_2 = 1.5\pi$. The heavy solid curve with $\alpha_2 = 1.445\pi$ has 0 curvatures at end points and belongs to $H_n(P_1, P_2; l)$ and $H_{m,1}(P_1, P_2; \alpha_1; l)$.

Next, we present *open* minimal-energy spline segments.

We compute a family of various-length *open* clamped minimal-energy spline segments with the constraint sets $H_c(P_1, P_2; \alpha_1, \alpha_2; l)$, where $P_1 = (-1, 0)$, $P_2 = (1, 0)$, $\alpha_1 = \pi/2$, $\alpha_2 = -\pi/2$, and $l \in [\pi, 3\pi]$. Six open clamped minimal-energy spline segments are shown in Fig. 2.2a, with their corresponding energies listed in Table 2.2a. When $l = \pi$, the corresponding spline segment is a semi-circle. The solid curve with $l = 1.39\pi$, $\alpha_1 = \pi/2$, and $\alpha_2 = -\pi/2$ has 0 curvatures at end points.

If there are no end constraints, the curvatures at end point equal 0. We compute a family of various-length *open* natural minimal-energy spline segments with the constraint sets $H_n(P_1, P_2; l)$, where $P_1 = (-1, 0)$, $P_2 = (1, 0)$, and $l \in [\pi, 3\pi]$. Six open natural minimal-energy spline segments are shown in Fig. 2.2b, with their corresponding energies listed in Table 2.2b. The solid curve with $l = 1.39\pi$, $\kappa_1 = 0$, and $\kappa_2 = 0$ has end angles $\alpha_1 = \pi/2$ and $\alpha_2 = -\pi/2$.

We also compute a family of various-length *open* mixed minimal-energy spline segments with the constraint sets $H_{m,1}(P_1, P_2; \alpha_1; l)$, where $P_1 = (-1, 0)$, $P_2 = (1, 0)$, $\alpha_1 = \pi/2$, and $l \in [\pi, 3\pi]$. Six open mixed minimal-energy spline segments are shown in Fig. 2.2c, with their corresponding energies listed in Table 2.2c. The solid curve with $l = 1.39\pi$, $\alpha_1 = \pi/2$, and $\kappa_2 = 0$ has end curvature $\kappa_1 = 0$ and end angle $\alpha_2 = -\pi/2$.

Now, we present minimal-energy splines.

Table 2.1

| $P_1$ | $P_2$ | $\alpha_1$ | $\alpha_2$ | length | energy |
|-------|-------|------------|------------|--------|--------|
| (0,0) | (0,0) | 0 | $1.0\pi$   | $\pi$ | 5.8340 |
| (0,0) | (0,0) | 0 | $1.1\pi$   | $\pi$ | 5.2961 |
| (0,0) | (0,0) | 0 | $1.2\pi$   | $\pi$ | 4.8933 |
| (0,0) | (0,0) | 0 | $1.3\pi$   | $\pi$ | 4.6254 |
| (0,0) | (0,0) | 0 | $1.4\pi$   | $\pi$ | 4.4911 |
| (0,0) | (0,0) | 0 | $1.445\pi$ | $\pi$ | 4.4722 |
| (0,0) | (0,0) | 0 | $1.5\pi$   | $\pi$ | 4.4881 |
| (0,0) | (0,0) | 0 | $1.6\pi$   | $\pi$ | 4.6129 |
| (0,0) | (0,0) | 0 | $1.7\pi$   | $\pi$ | 4.8611 |
| (0,0) | (0,0) | 0 | $1.8\pi$   | $\pi$ | 5.2269 |
| (0,0) | (0,0) | 0 | $1.9\pi$   | $\pi$ | 5.7035 |
| (0,0) | (0,0) | 0 | $2.0\pi$   | $\pi$ | 6.2832 |

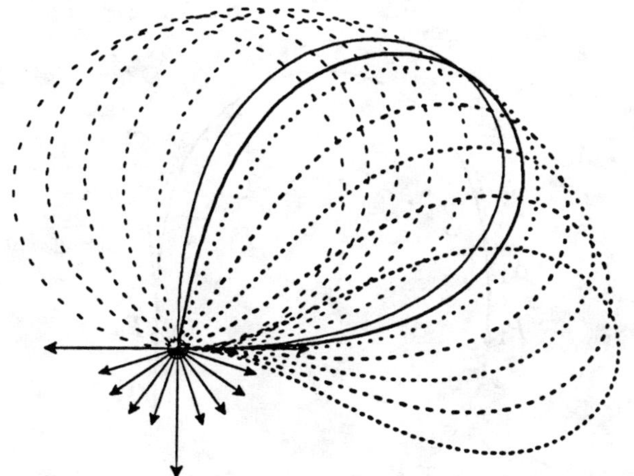

$P_1 = (0,0)$, $P_2 = (0,0)$, $\alpha_1 = 0$, $\alpha_2 \in [\pi, 2\pi]$, length $= \pi$

Fig. 2.1.  *Closed minimal-energy spline segments.*

TABLE 2.2a

| $P_1$ | $P_2$ | $\alpha_1$ | $\alpha_2$ | *length* | *energy* |
|:---:|:---:|:---:|:---:|:---:|:---:|
| $(-1,0)$ | $(1,0)$ | $\pi/2$ | $-\pi/2$ | $1.00\pi$ | 1.5708 |
| $(-1,0)$ | $(1,0)$ | $\pi/2$ | $-\pi/2$ | $1.39\pi$ | 1.4354 |
| $(-1,0)$ | $(1,0)$ | $\pi/2$ | $-\pi/2$ | $1.50\pi$ | 1.4347 |
| $(-1,0)$ | $(1,0)$ | $\pi/2$ | $-\pi/2$ | $2.00\pi$ | 1.3899 |
| $(-1,0)$ | $(1,0)$ | $\pi/2$ | $-\pi/2$ | $2.50\pi$ | 1.2994 |
| $(-1,0)$ | $(1,0)$ | $\pi/2$ | $-\pi/2$ | $3.00\pi$ | 1.1996 |

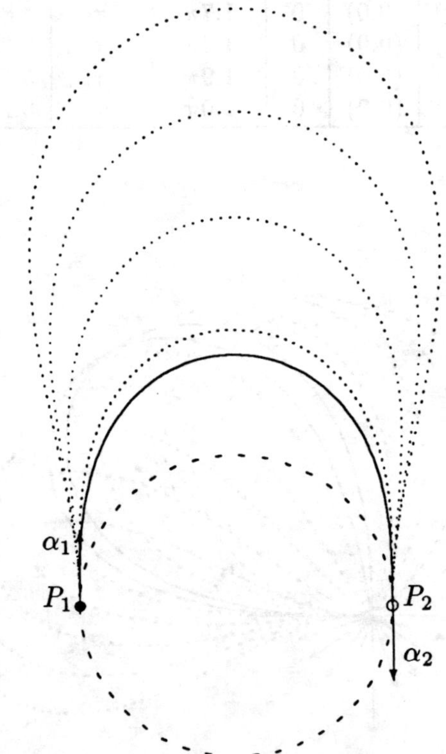

$P_1 = (-1,0),\ P_2 = (1,0),\ \alpha_1 = 0.5\pi,\ \alpha_2 = -0.5\pi,\ \text{length} \in [\pi, 3\pi]$

FIG. 2.2a. *Open clamped minimal-energy spline segments.*

TABLE 2.2b

| $P_1$ | $P_2$ | $\kappa_1$ | $\kappa_2$ | *length* | *energy* |
|:---:|:---:|:---:|:---:|:---:|:---:|
| $(-1,0)$ | $(1,0)$ | 0 | 0 | $1.00\pi$ | 1.2618 |
| $(-1,0)$ | $(1,0)$ | 0 | 0 | $1.39\pi$ | 1.4354 |
| $(-1,0)$ | $(1,0)$ | 0 | 0 | $1.50\pi$ | 1.4290 |
| $(-1,0)$ | $(1,0)$ | 0 | 0 | $2.00\pi$ | 1.3202 |
| $(-1,0)$ | $(1,0)$ | 0 | 0 | $2.50\pi$ | 1.1844 |
| $(-1,0)$ | $(1,0)$ | 0 | 0 | $3.00\pi$ | 1.0618 |

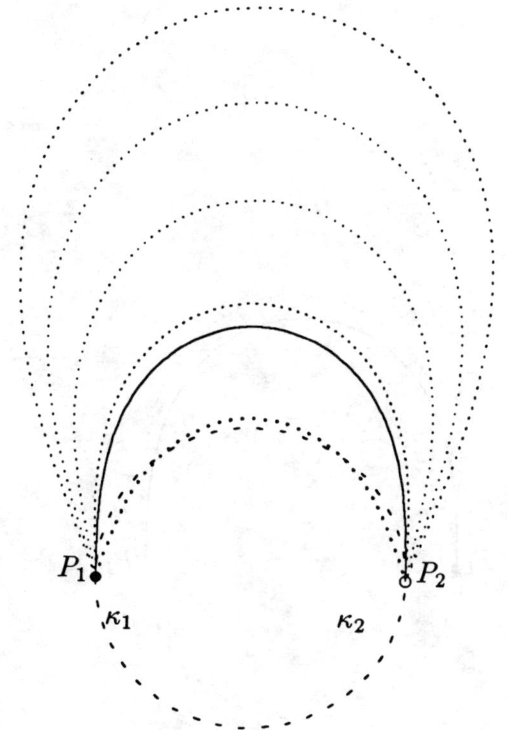


$$P_1 = (-1,0),\ P_2 = (1,0),\ \kappa_1 = 0,\ \kappa_2 = 0,\ \text{length} \in [\pi, 3\pi]$$

FIG. 2.2b. *Open natural minimal-energy spline segments.*

TABLE 2.2c

| $P_1$ | $P_2$ | $\alpha_1$ | $\kappa_2$ | *length* | *energy* |
|-------|-------|-----------|-----------|----------|----------|
| $(-1,0)$ | $(1,0)$ | $\pi/2$ | 0 | $1.00\pi$ | 1.3496 |
| $(-1,0)$ | $(1,0)$ | $\pi/2$ | 0 | $1.39\pi$ | 1.4354 |
| $(-1,0)$ | $(1,0)$ | $\pi/2$ | 0 | $1.50\pi$ | 1.4303 |
| $(-1,0)$ | $(1,0)$ | $\pi/2$ | 0 | $2.00\pi$ | 1.3355 |
| $(-1,0)$ | $(1,0)$ | $\pi/2$ | 0 | $2.50\pi$ | 1.2070 |
| $(-1,0)$ | $(1,0)$ | $\pi/2$ | 0 | $3.00\pi$ | 1.0862 |

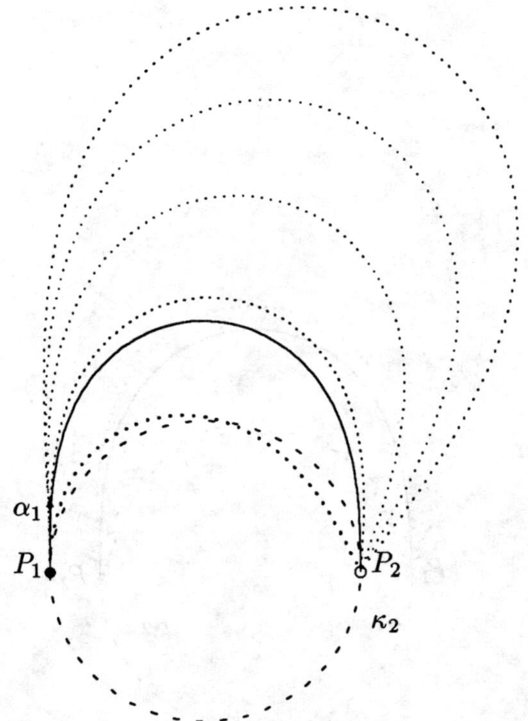


$$P_1 = (-1,0),\ P_2 = (1,0),\ \alpha_1 = 0.5\pi,\ \kappa_2 = 0,\ \text{length} \in [\pi, 3\pi]$$

FIG. 2.2c. *Open mixed minimal-energy splines.*

TABLE 2.3a

| $P_0$ | $P_1$ | $P_2$ | $P_3$ | $\alpha_0$ | $\alpha_3$ | $l_1$ | $l_2$ | $l_3$ | energy |
|---|---|---|---|---|---|---|---|---|---|
| $(0,1)$ | $(\frac{\sqrt{3}}{2}, -\frac{1}{2})$ | $(-\frac{\sqrt{3}}{2}, -\frac{1}{2})$ | $(0,1)$ | $0.50\pi$ | $-2.50\pi$ | $2.8084$ | $1.8564$ | $2.8084$ | $8.1294$ |
| $(0,1)$ | $(\frac{\sqrt{3}}{2}, -\frac{1}{2})$ | $(-\frac{\sqrt{3}}{2}, -\frac{1}{2})$ | $(0,1)$ | $0.25\pi$ | $-2.25\pi$ | $2.8084$ | $1.8564$ | $2.8084$ | $4.3807$ |
| $(0,1)$ | $(\frac{\sqrt{3}}{2}, -\frac{1}{2})$ | $(-\frac{\sqrt{3}}{2}, -\frac{1}{2})$ | $(0,1)$ | $0.0455\pi$ | $-2.0455\pi$ | $2.8084$ | $1.8564$ | $2.8084$ | $3.4691$ |
| $(0,1)$ | $(\frac{\sqrt{3}}{2}, -\frac{1}{2})$ | $(-\frac{\sqrt{3}}{2}, -\frac{1}{2})$ | $(0,1)$ | $0$ | $-2\pi$ | $2\pi/3$ | $2\pi/3$ | $2\pi/3$ | $3.1416$ |
| $(0,1)$ | $(\frac{\sqrt{3}}{2}, -\frac{1}{2})$ | $(-\frac{\sqrt{3}}{2}, -\frac{1}{2})$ | $(0,1)$ | $-0.25\pi$ | $-1.75\pi$ | $2.8084$ | $1.8564$ | $2.8084$ | $5.2553$ |
| $(0,1)$ | $(\frac{\sqrt{3}}{2}, -\frac{1}{2})$ | $(-\frac{\sqrt{3}}{2}, -\frac{1}{2})$ | $(0,1)$ | $-0.50\pi$ | $-1.50\pi$ | $2.8084$ | $1.8564$ | $2.8084$ | $8.7358$ |

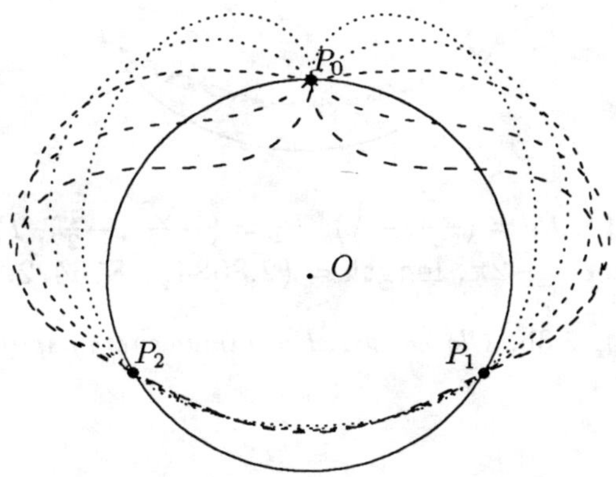

$$P_0 = (0,1),\ P_1 = (\tfrac{\sqrt{3}}{2}, -\tfrac{1}{2}),\ P_2 = (-\tfrac{\sqrt{3}}{2}, -\tfrac{1}{2}),\ P_3 = (0,1),$$
$$\alpha_0 + \alpha_3 = -2\pi,\ \text{length} = (2.8084, 1.8564, 2.8084)$$

FIG. 2.3a. *Closed clamped minimal-energy splines.*

TABLE 2.3b

| $P_0$ | $P_1$ | $P_2$ | $P_3$ | $\alpha_0$ | $l_1$ | $l_2$ | $l_3$ | energy |
|---|---|---|---|---|---|---|---|---|
| $(0,1)$ | $(\frac{\sqrt{3}}{2},-\frac{1}{2})$ | $(-\frac{\sqrt{3}}{2},-\frac{1}{2})$ | $(0,1)$ | $0.50\pi$ | 2.8084 | 1.8564 | 2.8084 | 5.7548 |
| $(0,1)$ | $(\frac{\sqrt{3}}{2},-\frac{1}{2})$ | $(-\frac{\sqrt{3}}{2},-\frac{1}{2})$ | $(0,1)$ | $0.25\pi$ | 2.8084 | 1.8564 | 2.8084 | 3.9169 |
| $(0,1)$ | $(\frac{\sqrt{3}}{2},-\frac{1}{2})$ | $(-\frac{\sqrt{3}}{2},-\frac{1}{2})$ | $(0,1)$ | $0.0455\pi$ | 2.8084 | 1.8564 | 2.8084 | 3.4691 |
| $(0,1)$ | $(\frac{\sqrt{3}}{2},-\frac{1}{2})$ | $(-\frac{\sqrt{3}}{2},-\frac{1}{2})$ | $(0,1)$ | $0$ | $2\pi/3$ | $2\pi/3$ | $2\pi/3$ | 3.1416 |
| $(0,1)$ | $(\frac{\sqrt{3}}{2},-\frac{1}{2})$ | $(-\frac{\sqrt{3}}{2},-\frac{1}{2})$ | $(0,1)$ | $-0.25\pi$ | 2.8084 | 1.8564 | 2.8084 | 4.3537 |
| $(0,1)$ | $(\frac{\sqrt{3}}{2},-\frac{1}{2})$ | $(-\frac{\sqrt{3}}{2},-\frac{1}{2})$ | $(0,1)$ | $-0.50\pi$ | 2.8084 | 1.8564 | 2.8084 | 6.0809 |

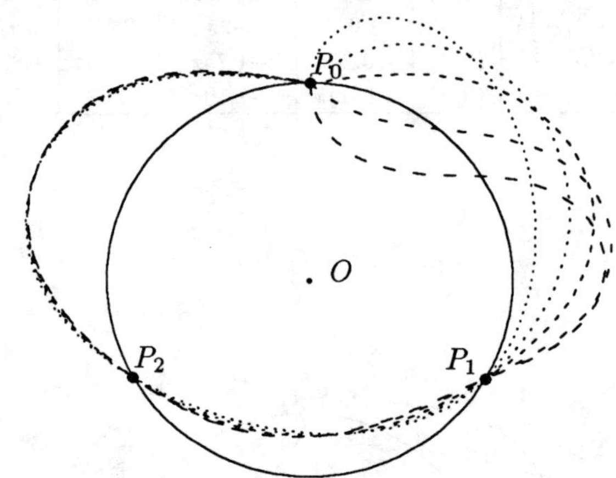


$$P_0 = (0,1), \; P_1 = (\tfrac{\sqrt{3}}{2},-\tfrac{1}{2}), \; P_2 = (-\tfrac{\sqrt{3}}{2},-\tfrac{1}{2}), \; P_3 = (0,1),$$
$$\alpha_0 + \alpha_3 = -2\pi, \; \text{length} = (2.8084, 1.8564, 2.8084)$$

FIG. 2.3b. *Closed mixed minimal-energy splines.*

We compute a family of four-point fixed-length *closed* clamped minimal-energy splines with the constraint sets $H_c(P_0, P_1, P_2, P_3; \alpha_0, \alpha_3; l_1, l_2, l_3)$, where $P_0 = P_3 = (0,1)$, $P_1 = (\sqrt{3}/2, -1/2)$, $P_2 = (-\sqrt{3}/2, -1/2)$, $\alpha_0 \in [-0.5\pi, 0.5\pi]$, $\alpha_0 + \alpha_3 = -2\pi$, $l_1 = 2.8084$, $l_2 = 1.8564$, and $l_3 = 2.8084$. The five closed clamped minimal-energy splines are shown in Fig. 2.3a, with their corresponding energies listed in Table 2.3a. The reference circle is obtained with $\alpha_0 = 0$, $\alpha_3 = -2\pi$, and $l_1 = l_2 = l_3 = 2\pi/3$. The curve with $\alpha_0 = 0.04545\pi$ and $\alpha_3 = -2.04545\pi$ has 0 curvatures at end points $P_0$ and $P_3$.

We also compute a family of four-point fixed-length *closed* mixed minimal-energy splines with the constraint sets $H_{m,1}(P_0, P_1, P_2, P_3; \alpha_0; l_1, l_2, l_3)$, where $P_0 = P_3 = (0,1)$, $P_1 = (\sqrt{3}/2, -1/2)$, $P_2 = (-\sqrt{3}/2, -1/2)$, $\alpha_0 \in [-0.5\pi, 0.5\pi]$, $l_1 = 2.8084$, $l_2 = 1.8564$, and $l_3 = 2.8084$. The five closed mixed minimal-energy splines are shown in Fig. 2.3b, with their corresponding

TABLE 2.4

| $P_0$ | $P_1$ | $P_2$ | $P_3$ | $\alpha_0$ | $\alpha_3$ | $l_1$ | $l_2$ | $l_3$ | energy |
|---|---|---|---|---|---|---|---|---|---|
| $(-2,0)$ | $(0,0)$ | $(2,0)$ | $(4,0)$ | $\pi/4$ | $-\pi/2$ | 3.1416 | 3.1416 | 3.1416 | 4.0641 |
| $(-2,0)$ | $(0,0)$ | $(2,0)$ | $(4,0)$ | $\pi/4$ | | 3.6832 | 3.1416 | 3.6832 | 4.4227 |
| $(-2,0)$ | $(0,0)$ | $(2,0)$ | $(4,0)$ | | | 3.1416 | 3.6832 | 3.1416 | 3.9360 |

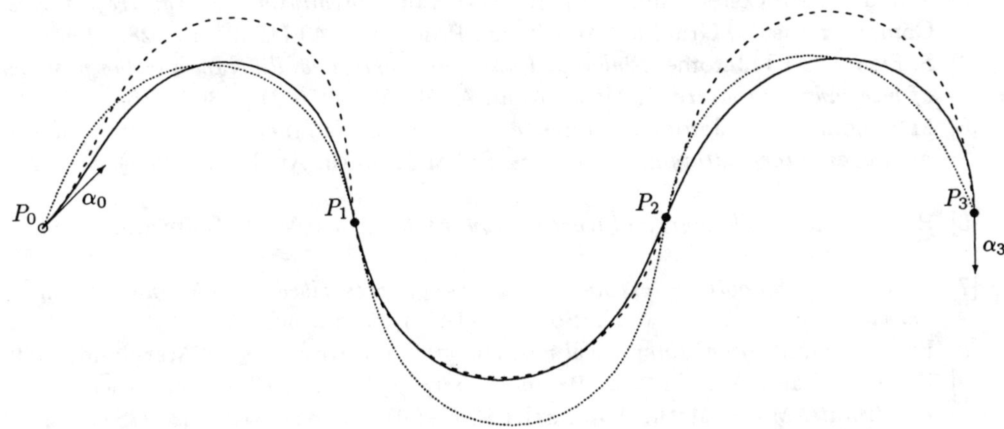


$$P_0 = (-2,0),\ P_1 = (0,0),\ P_2 = (2,0),\ P_3 = (4,0)$$
$$\alpha_0 = 0.25\pi,\ \alpha_3 = -0.5\pi,\ \text{length} \in [3.1416, 3.6832]$$

FIG. 2.4. *Open minimal-energy splines.*

energies listed in Table 2.3b. The reference circle is obtained with $\alpha_0 = 0$, $\alpha_3 = -2\pi$, and $l_1 = l_2 = l_3 = 2\pi/3$. The curve with $\alpha_0 = 0.04545\pi$ has 0 curvatures at end points $P_0$ and $P_3$.

Note that the four-point fixed-length *closed* natural minimal-energy spline with the constraint set $H_n(P_0, P_1, P_2, P_3; l_1, l_2, l_3)$, where $P_0 = P_3 = (0,1)$, $P_1 = (\sqrt{3}/2, -1/2)$, $P_2 = (-\sqrt{3}/2, -1/2)$, $l_1 = 2.8084$, $l_2 = 1.8564$, and $l_3 = 2.8084$, has end angles $\alpha_0 = 0.04545\pi$ and $\alpha_3 = -2.04545\pi$ and is the same curve with $H_c(P_0, P_1, P_2, P_3; \alpha_0, \alpha_3; l_1, l_2, l_3)$ or $H_{m,1}(P_0, P_1, P_2, P_3; \alpha_0; l_1, l_2, l_3)$.

Finally, we compute four-point various-length *open* clamped, natural, and mixed minimal-energy splines with the constraint sets $H_c(P_0, P_1, P_2, P_3;\ \alpha_0, \alpha_3; l_1, l_2, l_3)$, $H_n(P_0, P_1, P_2, P_3; l_1, l_2, l_3)$, and $H_{m,1}(P_0, P_1, P_2, P_3; \alpha_0; l_1, l_2, l_3)$, respectively, where $P_0 = (-2,0)$, $P_1 = (0,0)$, $P_2 = (2,0)$, $P_3 = (4,0)$, $\alpha_0 = \pi/4$, $\alpha_3 = -\pi/2$, and $l_1 = 4.2249$, $l_2 = 3.1416$, and $l_3 = 3.6832$. The three open minimal-energy splines are shown in Fig. 2.4, with their corresponding energies listed in Table 2.4. The solid, dotted, and dashed curves correspond to the clamped, natural, and mixed minimal-energy splines, respectively.

## References


[1] G. Birkhoff and C. R. de Boor, *Piecewise polynomial interpolation and approximation of functions*, Proceedings of General Motors Symposium of 1964, H. L. Garabedian, ed., Elsevier, Amsterdam, 1965, pp. 164–190.

[2] G. Birkhoff, H. Burchard, and D. Thomas, *Nonlinear Interpolation by Splines, Pseudospline, and Elastica*, Research Publication GMR–468, General Motors Corporation, Warren, Michigan, 1965.

[3] Alfred M. Bruckstein and Arun N. Netravali, *On minimal energy trajectories*, Computer Vision, Graphics, and Image Processing, 49 (1990), pp. 283–296.

[4] S. Fisher and J. Jerome, *Stable and unstable elastica equilibrium and the problem of minimum curvature*, J. Math. Anal. Appl., 53 (1976), pp. 367–376.

[5] M. Golomb and J. Jerome, *Equilibria of the curvature functional and manifolds of nonlinear interpolating spline curves*, SIAM J. Math. Anal., 13 (1982), pp. 421–458.

[6] B. K. P. Horn, *The curve of least energy*, ACM Trans. Math. Software, 9 (1983), pp. 441–469.

[7] J. Jerome, *Smooth interpolating curves of prescribed length and minimum curvature*, Proc. Amer. Math. Soc., 51 (1975), pp. 62–66.

[8] Emery Jou, *Minimal Energy Splines*, Dissertation, University of Maryland, 1989.

[9] Emery Jou and Weimin Han, *Minimal-energy splines: I. Plane curves with angle constraints*, Math. Meth. Appl. Sci., 13 (1990), pp. 351–372. Also CS-TR-2533 (UMIACS-TR-90-119), Comp. Sci. Dept., Univ. of Maryland, 1990.

[10] ——, *Minimal-energy splines: II. Mixed and natural plane curves*, CS-TR-2534 (UMIACS-TR-90-120), Comp. Sci. Dept., Univ. of Maryland, 1990.

[11] Weimin Han and Emery Jou, *On the computation of minimal-energy splines: Convergence analysis*, Appl. Math. Comp., 47 (1992), pp. 1–13; also CS-TR-2545 (UMIACS-TR-90-129), Comp. Sci. Dept., Univ. of Maryland, 1990.

[12] Emery Jou and Weimin Han, *Minimal-energy splines with point forces*, CS-TR-2546 (UMIACS-TR-90-130), Comp. Sci. Dept., Univ. of Maryland, 1990.

[13] Michael Kallay, *Plane curves of minimal energy*, ACM Trans. Math. Software, 12 (1986), pp. 219–222.

[14] E. H. Lee and G. E. Forsythe, *Variational study of nonlinear spline curves*, SIAM Rev., 15 (1975), pp. 120–133.

[15] A. E. H. Love, *A Treatise on the Mathematical Theory of Elasticity*, Dover, New York, 1944.

[16] Michael A. Malcolm, *On the computation of nonlinear spline functions*, SIAM J. Numer. Anal., 14 (1977), pp. 254–282.

[17] Even Mehlum, *Curve and Surface Fitting Based on Variational Criteria for Smoothness*, Central Institute for Industrial Research, Oslo, Norway, 1969.

[18] Guido Schulze, *Elastische Wege und nichtlineare Splines im CAGD*, Dissertation, Universitaet Kaiserslautern, 1990.

# Interval Weighted Tau-Splines

Dieter Lasser and Hans Hagen

## 3.1. Introduction

In 1974 Nielson described a piecewise polynomial alternative to splines under tension, the so-called $\nu$-splines, curvature continuous interpolating cubics [7]. In 1985 Hagen generalized Nielson's approach resulting in polynomials of degree $n = 2K - 1$, $K \geq 2$, $K \in I\!\!N$ [4]. Hagen's concept of *geometric spline curves* includes for $K = 2$ Nielson's $\nu$-splines and for $K = 3$ curvature and torsion continuous quintics, the $\tau$-splines. $\tau$-splines have been discussed in detail in [6]. Geometric spline curves due to Nielson and Hagen provide shape parameters having the characteristics of point weights. The idea of interval weights was illustrated in [10] and in [2] for cubics, and in [3] Foley generalized Nielson's $\nu$-splines to interval weighted $\nu$-splines supplied in any $\mathbf{X}(t_I)$ point weights $\nu_I$ and in addition for every curve segment $\mathbf{X}_I$, $t \in [t_I, t_{I+1}]$, interval weights $p_I$. The corresponding minimization principle yielding interval weighted geometric spline curves due to Hagen has been given in [6]; variational principle and continuity conditions have also been given in [5]. To recapitulate, *interval weighted geometric spline curves* are the solutions of the minimization of

$$\sum_{I=0}^{N-1} p_I \int_{t_I}^{t_{I+1}} \|\mathbf{X}^{(K)}(t)\|^2 \, dt \quad + \sum_{I=0}^{N} \sum_{L=1}^{K-1} \nu_{I,L} \|\mathbf{X}^{(L)}(t_I)\|^2 \, ,$$

$$(3.1) \qquad\qquad K \geq 2, \quad \nu_{I,L} \geq 0, \quad p_I > 0,$$

over the space

$$\mathcal{H}^K = \left\{ \mathbf{X} : \mathbf{X}^{(K)} \in \mathcal{L}^2[t_0, t_N], \ \mathbf{X}^{(K-1)} \text{ absolutely continuous on } [t_0, t_N] \right\},$$

satisfying the interpolation conditions

$$(3.2) \qquad\qquad I = 0, \cdots, N \qquad \mathbf{X}(t_I) = \mathbf{X}_I$$

and one of the following generalized end conditions $(L = K, \cdots, 2(K-1))$:

$$\text{(i)} \quad \mathbf{X}^{(2K-1-L)}(t_0) = \mathbf{X}_0^{(2K-1-L)}, \qquad \mathbf{X}^{(2K-1-L)}(t_N) = \mathbf{X}_N^{(2K-1-L)},$$

$$\text{(ii)} \quad p_0\,\mathbf{X}^{(L)}(t_0^+) - (-1)^{K-L}\,\nu_{0,2K-1-L}\,\mathbf{X}^{(2K-1-L)}(t_0) = 0\,,$$

$$(3.3) \quad p_{N-1}\,\mathbf{X}^{(L)}(t_N^-) + (-1)^{K-L}\,\nu_{N,2K-1-L}\,\mathbf{X}^{(2K-1-L)}(t_N) = 0\,,$$

$$\text{(iii)} \quad \mathbf{X}(t_0) = \mathbf{X}(t_N), \qquad \mathbf{X}^{(2K-1-L)}(t_0) = \mathbf{X}^{(2K-1-L)}(t_N)\,,$$

$$p_0\,\mathbf{X}^{(L)}(t_0^+) - p_{N-1}\,\mathbf{X}^{(L)}(t_N^-)$$
$$= (-1)^{K-L}\left(\nu_{0,2K-1-L} + \nu_{N,2K-1-L}\right)\mathbf{X}^{(2K-1-L)}(t_0)\,.$$

The solutions of (3.1) are piecewise polynomials of degree $n = 2K-1$, $K \geq 2$, $K \in I\!N$, fulfilling at any knot $t_I$ $(I = 1, \cdots, N-1)$ the continuity conditions

$$(3.4) \quad \begin{array}{ll} L = 0, \cdots, K-1 & \mathbf{X}^{(L)}(t_I^+) = \mathbf{X}^{(L)}(t_I^-) \\[2mm] L = K, \cdots, 2K-2 & \mathbf{X}^{(L)}(t_I^+) = \pi_I \mathbf{X}^{(L)}(t_I^-) \\[1mm] & \quad +(-1)^{K-L}\mu_{I,2K-L-1}\mathbf{X}^{(2K-L-1)}(t_I)\,, \end{array}$$

where $\pi_I = \frac{p_{I-1}}{p_I}$ and $\mu_{I,L} = \frac{\nu_{I,L}}{p_I}$. Thus ([6]), interval weighted geometric spline curves are in $\mathbf{X}(t_I)$ in general not curvature-, torsion-, etc., continuous. That is, they provide only a *weak kind* of geometric continuity, compared with definitions of geometric continuity based on the Frenet frame and differential geometric invariants (curvatures), $F^r$-continuity, or the concept of contact of order $r$, $G^r$-continuity (see, e.g., [5]). The reason we still are interested in interval weighted geometric splines is closely connected to the answer of the question: Do we somewhere really actually *use* the minimization property of interval weighted splines? The answer is yes, of course! Working with different kinds of geometric spline curves, $F^r$-, $G^r$-continuous splines and splines derived from a variational principle, it becomes clear that shape parameters have the important tension character only if they are defined by a variational principle. We say a shape parameter possesses the tension character if the interpolating curve gets straightened and possibly converges towards the polygon that interpolates the points, for increasing (or decreasing) shape parameter values. Now, this is exactly the case for geometric splines defined by a variational principle, like (interval weighted) $\nu$-splines and $\tau$-splines,[1] but this is not the case for $F^r$- or for $G^r$-continuous spline curves![2, 3]

Of practical interest are $K = 2$ and $K = 3$, yielding interval weighted $\nu$-splines and $\tau$-splines, respectively. The first have been addressed already

---

[1] This statement has to be taken with care, for it does not really have unlimited validity: The $p_I$ and $\nu_{I,L}$ have tension character only within certain ranges, as we will see in §3.3.3.

[2] Except for $r = 2$, curvature continuous cubics are equivalent to $\nu$-splines.

[3] Curves holding higher order geometic continuity properties as well as being equipped with shape parameters which have tension character are described in [9]. But this is at the expense of the polynomial nature, for Pottmann's curves are built up by a polynomial and an exponential part.

in [3]. The second, independent of us, in [8], for curve and surface definition, illustrates very nicely the influence of the tension parameters on the curve and surface shape. But no B-spline-Bézier representations have been given yet. During the last years the B-spline-Bézier representation of curves and surfaces became very popular and can be thought of as a de facto industry standard. Therefore, the first aim of this paper is to give a Bézier representation of interval weighted $\tau$-splines. The Bézier representation can be embedded in a B-spline representation of quintic spline curves, yielding to a Bézier-based B-spline representation of interval weighted $\tau$-splines. We also derive conversion equations between the Bézier and the Hermite-like representation. Furthermore, we discuss positivity conditions on the shape parameters of the Bézier representation and value ranges for the control point and interval weights, such that the shape parameters of the B-spline-Bézier representation are positive; this ensures such important things as the convex hull and the variation diminishing properties of the B-spline-Bézier representation.

Because we would like to find a Bézier and a B-spline representation of interval weighted $\tau$-splines, we first introduce the Bézier representation of segmented curves. In §3.3 we give a short discussion, especially a B-spline-Bézier representation, of Foley's interval weighted $\nu$-splines which is also still missing in the literature. They are the starting point for the development of interval weighted $\tau$-splines, which will be discussed in §3.4.

### 3.2. Bézier Representation of Segmented Curves

Let $\mathbf{X}(t)$ be a planar or spatial parameterized curve defined with respect to a partition of the domain space by *knots* $t_0 < t_1 < \cdots < t_N$. The parameter space segmentation induces a curve segmentation in segments $\mathbf{X}_I : [t_I, t_{I+1}] \to I\!\!R^d$ $(d =\in I\!\!N)$. A local parameter $u \in [0,1]$ can be introduced such that $I = 0, \cdots, N-1 : \mathbf{X}(t) = \mathbf{X}_I(u)$ for $t \in [t_I, t_{I+1}]$ by the linear interpolation of $t_I$ and $t_{I+1}$: $t = (1-u)\,t_I + u\,t_{I+1}$, where $u \in [0,1]$. The derivatives have to be calculated now by the chain rule, i.e.,

$$\mathbf{X}^{(r)}(t) \equiv \frac{d^r}{dt^r}\mathbf{X}(t) = \frac{d^r u}{dt^r}\frac{d^r}{du^r}\mathbf{X}_I(u) = \frac{1}{\Delta_I^r}\frac{d^r}{du^r}\mathbf{X}_I(u),$$

where $\Delta_I \equiv t_{I+1} - t_I$.

Now $\mathbf{X}(t)$ might be given in B-spline representation, that means

$$(3.5) \qquad\qquad \mathbf{X}(t) = \sum_I \mathbf{d}_I\, G_I^n(t),$$

where the control points $\mathbf{d}_I \in I\!\!R^d$ are often called *de Boor points*, and with local support *basis functions* $G_I^n(t)$. In this paper the construction of $\mathbf{X}(t)$ is based on the *Bézier representation*

$$(3.6) \qquad\qquad \mathbf{X}_I(u) = \sum_{k=0}^{n} \mathbf{b}_{nI+k}\, B_k^n(u)$$

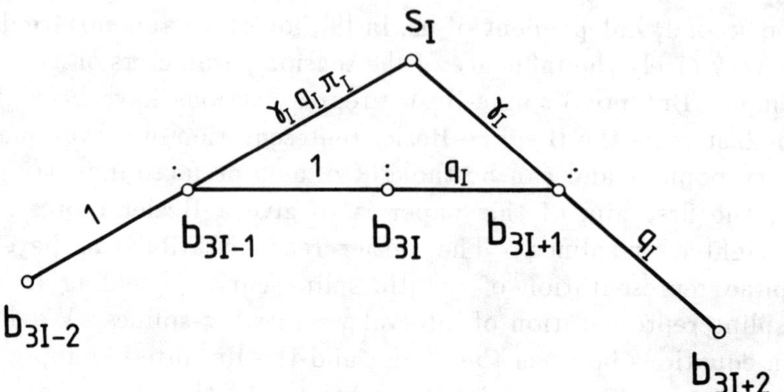


FIG. 3.1. *Bézier polygon for interval weighted $\nu$-splines.*

of the curve segments $\mathbf{X}_I(u)$. The control points $\mathbf{b}_{nI+k} \in I\!\!R^d$ are often called *Bézier points*. The $B_k^n(u)$ are the (ordinary) *Bernstein polynomials* of degree $n$ in $u$ (see, e.g., [5]).

## 3.3.  Interval Weighted Nu-Splines

Foley's interval weighted $\nu$-splines [3] are solutions of the minimization of (3.1) over the space $\mathcal{H}^K$, subject to the interpolation conditions (3.2) and one of the end conditions (3.3) fulfilling at any knot $t_I$ $(I = 1, \cdots, N-1)$ continuity conditions (3.4), for $K = 2$, i.e., $n = 3$.

### 3.3.1.  Bézier Representation.

A Bézier representation of interval weighted $\nu$-splines can be derived by substituting (3.6) into (3.4), for $K = 2$. We get as continuity conditions of the Bézier representation (see Fig. 3.1)

$$(3.7) \qquad \begin{aligned} (1 + q_I)\mathbf{b}_{3I} &= q_I\mathbf{b}_{3I-1} + \mathbf{b}_{3I+1}, \\ (1 + \gamma_I q_I \pi_I)\mathbf{b}_{3I-1} &= \gamma_I q_I \pi_I \mathbf{b}_{3I-2} + \mathbf{s}_I, \\ (\gamma_I + q_I)\mathbf{b}_{3I+1} &= q_I\mathbf{s}_I + \gamma_I\mathbf{b}_{3I+2}, \end{aligned}$$

where

$$(3.8) \qquad \gamma_I = \frac{1 + q_I}{1 + q_I\pi_I + \frac{\Delta_I}{2}\mu_I},$$

$q_I = \frac{\Delta_I}{\Delta_{I-1}}$, $I = 1, \cdots, N-1$, and $\mu_I \equiv \mu_{I,1}$. Equation (3.8) allows the evaluation of the $\gamma_I$s of the Bézier representation of an interpolating interval weighted $\nu$-spline, i.e., the evaluation of the $\gamma_I$s for given $\mu_I$-values. On the other hand, the $\mu_I$s of a Hermite-like representation for given $\gamma_I$s of a Bézier representation are determined by

$$(3.9) \qquad I = 1, \cdots, N-1 \qquad \mu_I = \frac{2}{\Delta_I}\left[(1 + q_I)\frac{1}{\gamma_I} - (1 + q_I\pi_I)\right].$$

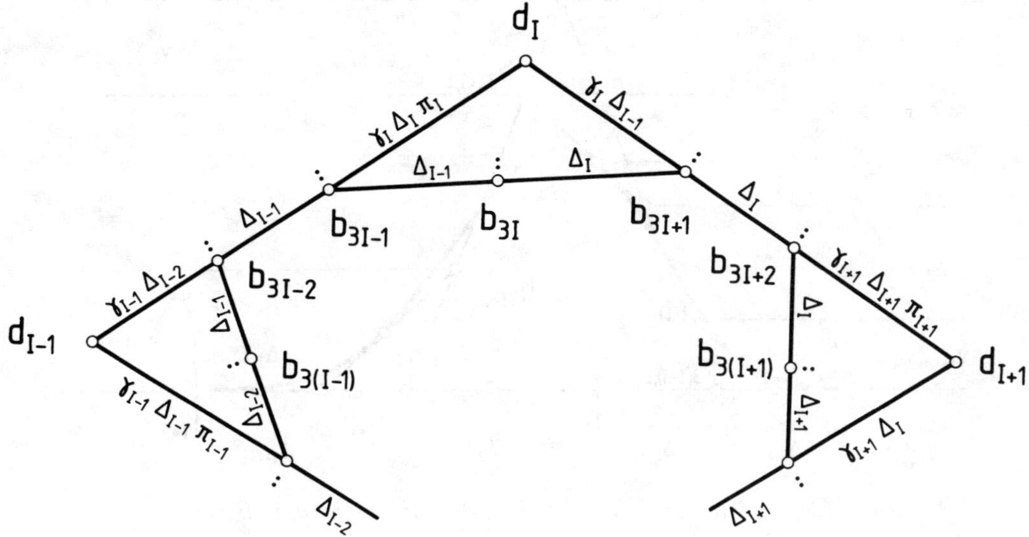


FIG. 3.2. *B-spline-Bézier polygon for interval weighted $\nu$-splines.*

**3.3.2. B-Spline Representation.** In the same way as for $C^2$-continuous cubics (see, e.g., [5]) we have for interval weighted $\nu$-splines (cf. Fig. 3.2)

$$
(3.10) \quad
\begin{aligned}
\Delta_{I,1}\mathbf{b}_{3I+1} &= (\Delta_I + \gamma_{I+1}\Delta_{I+1}\pi_{I+1})\mathbf{d}_I + \gamma_I\Delta_{I-1}\mathbf{d}_{I+1}, \\
\Delta_{I,1}\mathbf{b}_{3I+2} &= \gamma_{I+1}\Delta_{I+1}\pi_{I+1}\mathbf{d}_I + (\gamma_I\Delta_{I-1} + \Delta_I)\mathbf{d}_{I+1},
\end{aligned}
$$

where

$$
\Delta_{I,1} = \gamma_I\Delta_{I-1} + \Delta_I + \gamma_{I+1}\Delta_{I+1}\pi_{I+1}.
$$

Equations (3.7) and (3.10) allow the calculation of the Bézier points $\mathbf{b}_{3I+k}$ of an interval weighted $\nu$-spline in B-spline representation, if control points $\mathbf{d}_I$, corresponding knots $t_I$, interval weights $p_I$, and point weights $\nu_I$ are given.

Please note that a nonuniform knot sequence is possible, allowing multiple knots.

On the other hand, suppose we are given an interval weighted $\nu$-spline in Bézier representation, then the control points $\mathbf{d}_I$ of a B-spline representation follow from (3.7) and (3.10) with $\mathbf{d}_I = \mathbf{s}_I$.

The local support basis functions $G_I^3(t)$ of (3.5), for $n = 3$, are polynomials, and therefore can be given in Bézier form as segmented nonparametric Bézier curves. To construct the Bézier ordinates of the Bézier representation of $G_I^3(t)$, we consider the control polygon that is obtained by setting the de Boor ordinate $d_I = 1$, while setting all other de Boor ordinates $d_J = 0$ $(J \neq I)$, i.e., we single out the basis function $G_I^3(t)$ by the identity $G_I^3(t) = \sum_J \delta_{IJ} G_J^3(t)$, where $\delta_{IJ}$ is the *Kronecker delta*. Thus, the Bézier ordinates of the $G_I^3(t)$ are a simple consequence of (3.7) and (3.10) and are given by (see Fig. 3.3)

$$
\begin{aligned}
b_{3I-2,I} &= \tfrac{1}{\Delta_{I-1,1}}\gamma_{I-1}\Delta_{I-2}, & b_{3I-1,I} &= \tfrac{1}{\Delta_{I-1,1}},(\gamma_{I-1}\Delta_{I-2} + \Delta_{I-1}), \\
b_{3I+2,I} &= \tfrac{1}{\Delta_{I,1}}\gamma_{I+1}\Delta_{I+1}\pi_{I+1}, & b_{3I+1,I} &= \tfrac{1}{\Delta_{I,1}},(\Delta_I + \gamma_{I+1}\Delta_{I+1}\pi_{I+1}),
\end{aligned}
$$

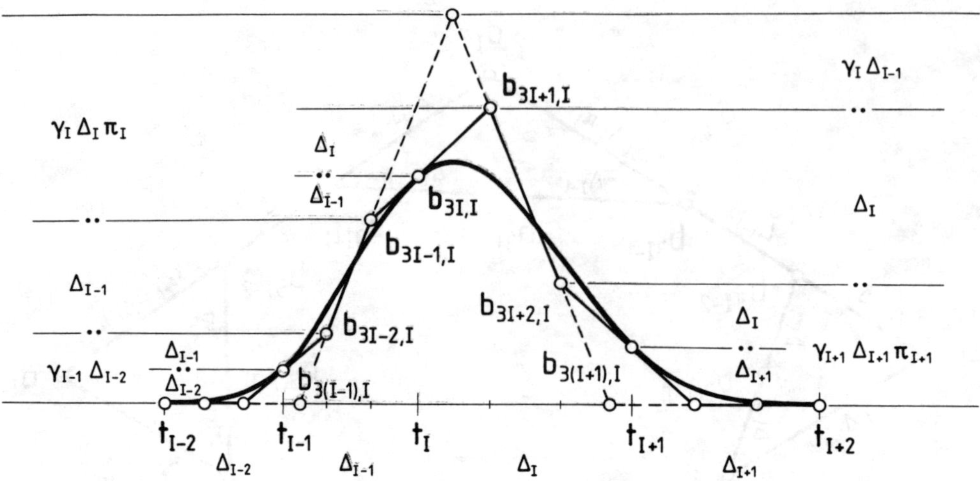



FIG. 3.3. *Bézier ordinates of the interval weighted $\nu$-B-splines.*

$$b_{3(I-1),I} = \frac{1}{\Delta_{I-1,2}}\Delta_{I-2}b_{3I-2,I},$$
$$b_{3I,I} = \frac{1}{\Delta_{I,2}}(\Delta_{I-1}b_{3I+1,I} + \Delta_I b_{3I-1,I}),$$
$$b_{3(I+1),I} = \frac{1}{\Delta_{I+1,2}}\Delta_{I+1}b_{3I+2,I},$$

where $\Delta_{I,2} = \Delta_{I-1} + \Delta_I$.

Note that the $G_I^3(t)$ form, by construction, a partition of unity.

### 3.3.3. Positivity of the Shape Parameters.

$p_I > 0$, $\nu_I \geq 0$ are the ranges for the tension parameters $p_I$ and $\nu_I$ covered by the minimum norm characterization of interval weighted $\nu$-splines [3]. These restrictions on $p_I$ and $\nu_I$ are equivalent to

$$I = 1, \cdots, N - 1 \qquad \gamma_I < \frac{1 + q_I}{1 + q_I \pi_I}.$$

On the other hand, if we require *positive shape parameters*, i.e., $\gamma_I > 0$, so that such important properties like the convex hull and the variation diminishing property of the B-spline-Bézier representation hold, (3.8) yields

$$I = 1, \cdots, N - 1 \qquad \mu_I > -\frac{2}{\Delta_I}(1 + q_I \pi_I).$$

Hence, not only positive tension values but $\nu_I$-values in the range given above guarantee positive shape parameters, if $p_I > 0$, i.e., $\pi_I > 0$.

### 3.4. Interval Weighted Tau-Splines

Interval weighted $\tau$-splines are solutions of the minimization of (3.1) over the space $\mathcal{H}^K$, subject to the interpolation conditions (3.2) and one of the end conditions (3.3) fulfilling at any knot $t_I$ ($I = 1, \cdots, N - 1$) continuity conditions (3.4), for $K = 3$, i.e., $n = 5$.

**3.4.1.  Bézier Representation.**   To find a Bézier representation of interval weighted $\tau$-splines we substitute (3.6) into (3.4), for $K = 3$, to obtain as continuity conditions for the Bézier representation (see Fig. 3.4)

$$(1 + q_I)\mathbf{b}_{5I} = q_I\mathbf{b}_{5I-1} + \mathbf{b}_{5I+1} ,$$

$$(1 + q_I)\mathbf{b}_{5I-1} = q_I\mathbf{b}_{5I-2} + \mathbf{s}_I ,$$
$$(1 + q_I)\mathbf{b}_{5I+1} = q_I\mathbf{s}_I + \mathbf{b}_{5I+2} ,$$

$$(1 + \delta_I q_I \pi_I)\mathbf{b}_{5I-2} = \delta_I q_I \pi_I \mathbf{b}_{5I-3} + \mathbf{e}_I^- ,$$
$$(\delta_I + \epsilon_I q_I)\mathbf{s}_I = \epsilon_I q_I \mathbf{e}_I^- + \delta_I \mathbf{e}_I^+ ,$$
$$(\epsilon_I + q_I)\mathbf{b}_{5I+2} = q_I \mathbf{e}_I^+ + \epsilon_I \mathbf{b}_{5I+3} ,$$

$$(1 + \rho_I q_I \pi_I)\mathbf{b}_{5I-3} = \rho_I q_I \pi_I \mathbf{b}_{5I-4} + \mathbf{f}_I^- ,$$
$$(\rho_I + \sigma_I q_I)\mathbf{e}_I^- = \sigma_I q_I \mathbf{f}_I^- + \rho_I \mathbf{t}_I ,$$
$$(\sigma_I + \tau_I q_I)\mathbf{e}_I^+ = \tau_I q_I \mathbf{t}_I + \sigma_I \mathbf{f}_I^+ ,$$
$$(\tau_I + q_I)\mathbf{b}_{5I+3} = q_I \mathbf{f}_I^+ + \tau_I \mathbf{b}_{5I+4} ,$$

(3.11)

where

$$(3.12)\quad \delta_I = \frac{(1 + q_I)^2}{1 + 2q_I\pi_I + q_I^2\pi_I + \frac{\Delta_I}{3}\mu_{I,2}} , \quad \epsilon_I = \frac{(1 + q_I)^2}{1 + 2q_I + q_I^2\pi_I + q_I\frac{\Delta_I}{3}\mu_{I,2}}$$

and

$$(3.13)\quad \rho_I = \frac{1}{3q_I\pi_I - R_I} , \quad \sigma_I = \frac{q_I^2\epsilon_I}{(T_I - 3)\epsilon_I + (T_I - S_I)q_I} , \quad \tau_I = \frac{q_I}{3 - T_I}$$

with

$$R_I = \frac{3q_I^4\pi_I^2 + 8q_I^3\pi_I^2 + 6q_I^2\pi_I^2 - \pi_I + \frac{\Delta_I^3}{24}\mu_{I,1}}{1 + 3q_I\pi_I + 3q_I^2\pi_I + q_I^3\pi_I + \frac{\Delta_I}{3}\mu_{I,2}} ,$$

$$S_I = \frac{2 + 4q_I - 4q_I^3\pi_I - 2q_I^4\pi_I + (1 - q_I)\frac{\Delta_I^3}{24}\mu_{I,1}}{1 + 3q_I + 3q_I^2\pi_I + q_I^3\pi_I + 2q_I\frac{\Delta_I}{3}\mu_{I,2}} ,$$

$$T_I = \frac{3 + 8q_I + 6q_I^2 - q_I^4\pi_I + q_I\frac{\Delta_I^3}{24}\mu_{I,1}}{1 + 3q_I + 3q_I^2 + q_I^3\pi_I + q_I^2\frac{\Delta_I}{3}\mu_{I,2}} .$$

Equations (3.12) and (3.13) allow the evaluation of the shape parameters of the Bézier representation of an interval weighted $\tau$-spline. The shape parameters are of course no longer independent of each other. This is obvious because three shape parameters, $\mu_{I,1}$, $\mu_{I,2}$, $\pi_I$, are given by (3.4), for $K = 3$, while six shape parameters, $\delta_I$, $\epsilon_I$, $\rho_I$, $\sigma_I$, $\tau_I$, $\pi_I$, are given by (3.11). The dependencies are

$$(3.14)\quad \delta_I(\epsilon_I) = \frac{(1 + q_I)^2\epsilon_I q_I}{(1 + q_I)^2 - \left[1 + 2q_I + q_I^2\pi_I - q_I(1 + 2q_I\pi_I + q_I^2\pi_I)\right]\epsilon_I}$$

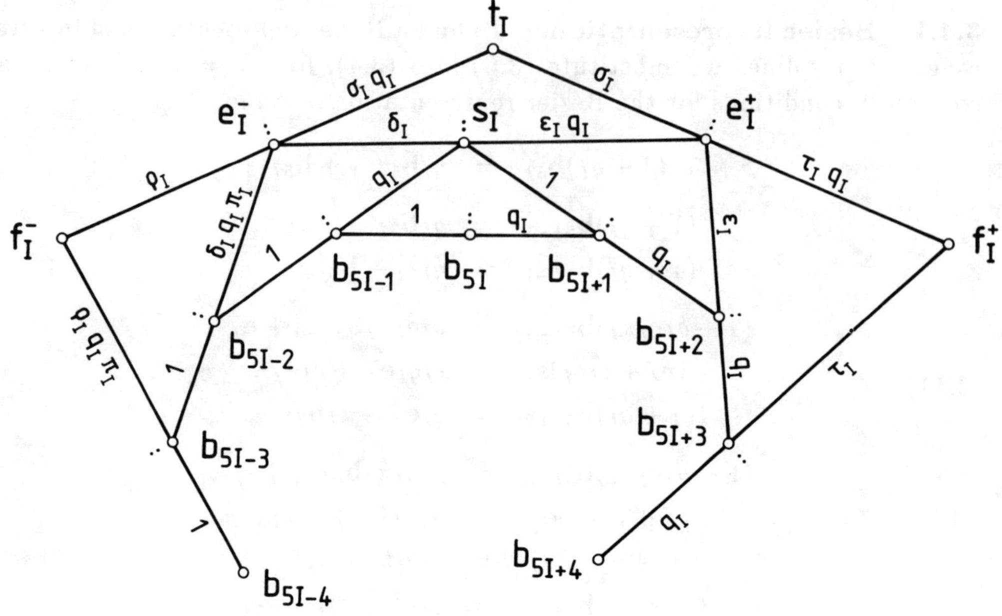


FIG. 3.4.  *Bézier polygon for interval weighted $\tau$-splines.*

and

$$(3.15) \qquad \rho_I(\sigma_I) = \frac{1}{\delta_I \pi_I} \frac{(1 + \delta_I q_I \pi_I)(\epsilon_I + \delta_I)\sigma_I}{3(1 + q_I)\sigma_I - (\delta_I + \epsilon_I q_I)},$$

$$(3.16) \qquad \sigma_I(\tau_I) = \frac{(\delta_I + q_I \epsilon_I)\epsilon_I \tau_I}{3(1 + q_I)\epsilon_I \tau_I - (\epsilon_I + q_I)(\epsilon_I + \delta_I)},$$

$$(3.17) \qquad \tau_I(\rho_I) = \frac{(\epsilon_I + q_I)\delta_I \rho_I \pi_I}{(1 + \delta_I q_I \pi_I)\epsilon_I}.$$

Hence, if an interpolating interval weighted $\tau$-spline is given, then the shape parameters of a corresponding Bézier representation have to be determined in such a way that they fulfill (3.14)–(3.17). And these equations are valid if the shape parameters are determined by (3.12) and (3.13). On the other hand, the shape parameters of a quintic Bézier spline curve have to fulfill conditions (3.14)–(3.17) to make the quintic Bézier spline be an interpolating interval weighted $\tau$-spline in Bézier representation. If the shape parameters of the Bézier representation are determined in this way, then the Bézier spline is equivalent to an interpolating interval weighted $\tau$-spline having the property of minimizing (3.1), for $K = 3$, and we can calculate point and interval weights by

$$(3.18) \qquad \mu_{I,2} = \frac{3}{\Delta_I}\left[(1 + q_I)^2 \frac{1}{\epsilon_I} - (1 + 2q_I + q_I^2 \pi_I)\right]\frac{1}{q_I},$$

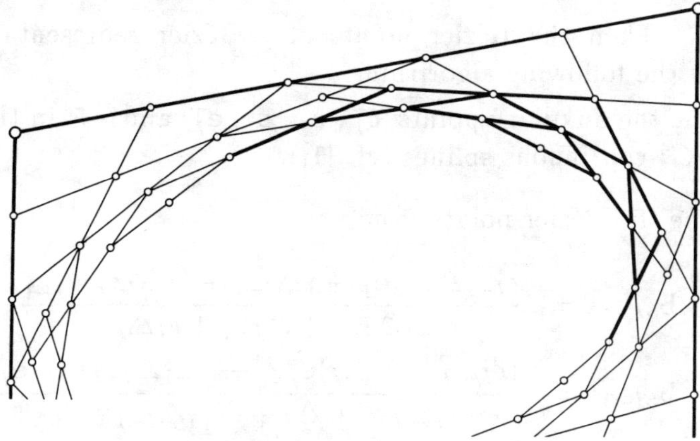

FIG. 3.5.  *A quintic lattice.*

$$\mu_{I,1} \;=\; \frac{24}{\triangle_I^3}\Bigg[\, 3 + 4q_I + q_I^4\pi_I$$

$$(3.19) \qquad\qquad\qquad -(1+q_I)^2\Big(6 + \Big(1+\tfrac{q_I}{\epsilon_I}\Big)\Big(\tfrac{q_I}{\tau_I}-3\Big)\Big)\Bigg]\tfrac{1}{q_I}\,.$$

### 3.4.2. B-Spline Representation.

For a B-spline representation the interval weighted $\tau$-spline construction of the Bézier representation (see Fig. 3.2) has to be embedded in a B-spline construction (see Fig. 3.5, the bold lines show the continuity construction of the Bézier representation of the two consecutive curve segments $\mathbf{X}_{I-1}$ and $\mathbf{X}_I$). This process yields the introduction of 10 auxiliary parameters, which depend on the shape parameters of the continuity construction. The first four dependencies are

$$\Theta_I \;=\; \frac{(\epsilon_{I-1}\triangle_{I-2} + \triangle_{I-1} + \delta_I\triangle_I\pi_I)\tau_{I-1}\delta_I\triangle_{I-1}}{(\epsilon_{I-1}\triangle_{I-2} + \triangle_{I-1})(\triangle_I + \triangle_{I-1}) - \tau_{I-1}\delta_I\triangle_{I-2}\triangle_I\pi_I},$$

$$\overline{\Theta}_I \;=\; \frac{(\epsilon_I\triangle_{I-1} + \triangle_I + \delta_{I+1}\triangle_{I+1}\pi_{I+1})\epsilon_I\rho_{I+1}\triangle_I}{(\triangle_{I-1} + \triangle_I)(\delta_{I+1}\triangle_{I+1}\pi_{I+1} + \triangle_I) - \rho_{I+1}\epsilon_I\triangle_{I-1}\triangle_{I+1}\pi_{I+1}},$$

$$\overline{\Lambda}_I \;=\; \frac{(\tau_I\triangle_{I-1} + \triangle_I + \triangle_{I+1})\tau_I\delta_{I+1}\triangle_I\pi_{I+1}}{(\epsilon_I\triangle_{I-1} + \triangle_I)(\triangle_{I+1} + \triangle_I) - \tau_I\delta_{I+1}\triangle_{I-1}\triangle_{I+1}\pi_{I+1}},$$

$$\Phi_I \;=\; \frac{(\triangle_{I-2} + \triangle_{I-1} + \rho_I\triangle_I\pi_I)\epsilon_{I-1}\rho_I\triangle_{I-1}}{(\triangle_{I-2} + \triangle_{I-1})(\delta_I\triangle_I\pi_I + \triangle_{I-1}) - \rho_I\epsilon_{I-1}\triangle_{I-2}\triangle_I\pi_I}.$$

while the auxiliary parameters $\Lambda_I$, $\overline{\Phi}_I$, $\Psi_I$, $\overline{\Psi}_I$, $\overline{\Omega}_I$, and $\Omega_I$ are given like for $GC^4$-continuous splines (cf. [1]).

Suppose that we are given control points $\mathbf{d}_I$, corresponding knots $t_I$, interval weights $p_I$ and point weights $\nu_{I,1}$ and $\nu_{I,2}$, and shape parameters $\delta_I$, $\epsilon_I$, $\rho_I$, $\sigma_I$, $\tau_I$, and thus the 10 auxiliary parameters of the B-spline representation. Then the Bézier points of a Bézier representation can be constructed by the following algorithm:

- Determine the auxiliary points $\mathbf{c}_I^-$, $\mathbf{e}_I^-$, $\mathbf{s}_I$, $\mathbf{e}_I^+$ and $\mathbf{c}_I^+$ in the same way as for $GC^4$-continuous splines (cf. [1]).

- Determine the Bézier points from

$$\mathbf{b}_{5I-3} = \frac{\epsilon_{I-1}\triangle_{I-2}\mathbf{e}_I^- + (\triangle_{I-1} + \delta_I\triangle_I\pi_I)\mathbf{e}_{I-1}^+}{\epsilon_{I-1}\triangle_{I-2} + \triangle_{I-1} + \delta_I\triangle_I\pi_I},$$

$$\mathbf{b}_{5I+3} = \frac{(\epsilon_I\triangle_{I-1} + \triangle_I)\mathbf{e}_{I+1}^- + \delta_{I+1}\triangle_{I+1}\pi_{I+1}\mathbf{e}_I^+}{\epsilon_I\triangle_{I-1} + \triangle_I + \delta_{I+1}\triangle_{I+1}\pi_{I+1}},$$

$$\mathbf{b}_{5I-2} = \frac{(\epsilon_{I-1}\triangle_{I-2} + \triangle_{I-1})\mathbf{e}_I^- + \delta_I\triangle_I\pi_I\mathbf{e}_{I-1}^+}{\epsilon_{I-1}\triangle_{I-2} + \triangle_{I-1} + \delta_I\triangle_I\pi_I},$$

$$\mathbf{b}_{5I+2} = \frac{\epsilon_I\triangle_{I-1}\mathbf{e}_{I+1}^- + (\triangle_I + \delta_{I+1}\triangle_{I+1}\pi_{I+1})\mathbf{e}_I^+}{\epsilon_I\triangle_{I-1} + \triangle_I + \delta_{I+1}\triangle_{I+1}\pi_{I+1}},$$

$$\mathbf{b}_{5I-1} = \frac{\triangle_{I-1}\mathbf{s}_I + \triangle_I\mathbf{b}_{5I-2}}{\triangle_{I-1} + \triangle_I},$$

$$\mathbf{b}_{5I+1} = \frac{\triangle_{I-1}\mathbf{b}_{5I+2} + \triangle_I\mathbf{s}_I}{\triangle_{I-1} + \triangle_I},$$

$$\mathbf{b}_{5I} = \frac{\triangle_{I-1}\mathbf{b}_{5I+1} + \triangle_I\mathbf{b}_{5I-1}}{\triangle_{I-1} + \triangle_I}.$$

On the other hand, suppose we are given an interval weighted $\tau$-spline in Bézier representation, then the auxiliary points $\mathbf{c}_{I+i,I}^\pm$, $\mathbf{e}_{I+i,I}^\pm$, $\mathbf{s}_{I+i,I}$ and the control points $\mathbf{d}_I$ of the B-spline representation follow as for $GC^4$-continuous quintics (see [1]).

Please note that a nonuniform knot sequence is possible again, thus allowing multiple knots.

A Bézier representation of the local support B-splines $G_I^5(t)$ can be derived in the same way as it was done in §3.3.2 for interval weighted $\nu$-splines, i.e., the 21 nonvanishing Bézier ordinates of $G_I^5(t)$ are given by

$$b_{5(I+1),I} = \frac{\triangle_I b_{5(I+1)+1,I}}{\triangle_I + \triangle_{I+1}},$$

$$b_{5(I+1)+1,I} = \frac{\triangle_I b_{5(I+1)+2,I}}{\triangle_I + \triangle_{I+1}},$$

$$b_{5(I+1)+2,I} = \frac{\epsilon_{I+1}\triangle_I e_{I+2,I}^-}{\epsilon_{I+1}\triangle_I + \triangle_{I+1} + \delta_{I+2}\triangle_{I+2}\pi_{I+2}},$$

$$b_{5(I+1)+3,I} = \frac{(\epsilon_{I+1}\triangle_I + \triangle_{I+1})e_{I+2,I}^-}{\epsilon_{I+1}\triangle_I + \triangle_{I+1} + \delta_{I+2}\triangle_{I+2}\pi_{I+2}},$$

$$b_{5(I+1)+4,I} = \frac{\triangle_{I+1}s_{I+2,I} + \triangle_{I+2}b_{5(I+1)+3,I}}{\triangle_{I+1} + \triangle_{I+2}},$$

$$b_{5(I+2),I} = \frac{\triangle_{I+1} b_{5(I+2)+1,I} + \triangle_{I+2} b_{5(I+1)+4,I}}{\triangle_{I+1} + \triangle_{I+2}},$$

$$b_{5(I+2)+1,I} = \frac{\triangle_{I+1} b_{5(I+2)+2,I} + \triangle_{I+2} s_{I+2,I}}{\triangle_{I+1} + \triangle_{I+2}},$$

$$b_{5(I+2)+2,I} = \frac{\epsilon_{I+2}\triangle_{I+1} e^-_{I+3,I} + (\triangle_{I+2} + \delta_{I+3}\triangle_{I+3}\pi_{I+3}) e^+_{I+2,I}}{\epsilon_{I+2}\triangle_{I+1} + \triangle_{I+2} + \delta_{I+3}\triangle_{I+3}\pi_{I+3}},$$

$$b_{5(I+2)+3,I} = \frac{(\epsilon_{I+2}\triangle_{I+1} + \triangle_{I+2}) e^-_{I+3,I} + \delta_{I+3}\triangle_{I+3}\pi_{I+3} e^+_{I+2,I}}{\epsilon_{I+2}\triangle_{I+1} + \triangle_{I+2} + \delta_{I+3}\triangle_{I+3}\pi_{I+3}},$$

$$b_{5(I+2)+4,I} = \frac{\triangle_{I+2} s_{I+3,I} + \triangle_{I+3} b_{5(I+2)+3,I}}{\triangle_{I+2} + \triangle_{I+3}},$$

$$b_{5(I+3),I} = \frac{\triangle_{I+2} b_{5(I+3)+1,I} + \triangle_{I+3} b_{5(I+2)+4,I}}{\triangle_{I+2} + \triangle_{I+3}},$$

$$b_{5(I+3)+1,I} = \frac{\triangle_{I+2} b_{5(I+3)+2,I} + \triangle_{I+3} s_{I+3,I}}{\triangle_{I+2} + \triangle_{I+3}},$$

$$b_{5(I+3)+2,I} = \frac{\epsilon_{I+3}\triangle_{I+2} e^-_{I+4,I} + (\triangle_{I+3} + \delta_{I+4}\triangle_{I+4}\pi_{I+4}) e^+_{I+3,I}}{\epsilon_{I+3}\triangle_{I+2} + \triangle_{I+3} + \delta_{I+4}\triangle_{I+4}\pi_{I+4}},$$

$$b_{5(I+3)+3,I} = \frac{(\epsilon_{I+3}\triangle_{I+2} + \triangle_{I+3}) e^-_{I+4,I} + \delta_{I+4}\triangle_{I+4}\pi_{I+4} e^+_{I+3,I}}{\epsilon_{I+3}\triangle_{I+2} + \triangle_{I+3} + \delta_{I+4}\triangle_{I+4}\pi_{I+4}},$$

$$b_{5(I+3)+4,I} = \frac{\triangle_{I+3} s_{I+4,I} + \triangle_{I+4} b_{5(I+3)+3,I}}{\triangle_{I+3} + \triangle_{I+4}},$$

$$b_{5(I+4),I} = \frac{\triangle_{I+3} b_{5(I+4)+1,I} + \triangle_{I+4} b_{5(I+3)+4,I}}{\triangle_{I+3} + \triangle_{I+4}},$$

$$b_{5(I+4)+1,I} = \frac{\triangle_{I+3} b_{5(I+4)+2,I} + \triangle_{I+4} s_{I+4,I}}{\triangle_{I+3} + \triangle_{I+4}},$$

$$b_{5(I+4)+2,I} = \frac{(\triangle_{I+4} + \delta_{I+5}\triangle_{I+5}\pi_{I+5}) e^+_{I+4,I}}{\epsilon_{I+4}\triangle_{I+3} + \triangle_{I+4} + \delta_{I+5}\triangle_{I+5}\pi_{I+5}},$$

$$b_{5(I+4)+3,I} = \frac{\delta_{I+5}\triangle_{I+5}\pi_{I+5} e^+_{I+4,I}}{\epsilon_{I+4}\triangle_{I+3} + \triangle_{I+4} + \delta_{I+5}\triangle_{I+5}\pi_{I+5}},$$

$$b_{5(I+4)+4,I} = \frac{\triangle_{I+5} b_{5(I+4)+3,I}}{\triangle_{I+4} + \triangle_{I+5}},$$

$$b_{5(I+5),I} = \frac{\triangle_{I+5} b_{5(I+4)+4,I}}{\triangle_{I+4} + \triangle_{I+5}},$$

where the auxiliary points $c^\pm_{I+i,I}$, $e^\pm_{I+i,I}$, $s_{I+i,I}$ of $G^5_I(t)$ have to be calculated as for $GC^4$-continuous quintics (see [1]).

Note that the $G^5_I(t)$ form, by construction, a partition of unity.

### 3.4.3. Positivity of the Shape Parameters.

Properties like the convex hull and variation diminishing property of the B-spline representation can be guaranteed only if additional restrictions are satisfied, i.e., if all parameters involved in the B-spline construction are positive. And since the Bézier construction is part of the B-spline construction, it is clear that all the shape parameters of the Bézier construction have to be positive. Therefore, because $q_I > 0$, the dependencies (3.14)–(3.17) imply the following.

Because of (3.14), positivity of $\delta_I = \delta_I(\epsilon_I)$ is given if and only if

$$0 < \epsilon_I < \frac{1+q_I}{1-q_I^2\pi_I} \qquad \text{for} \quad q_I^2\pi_I < 1\,,$$

$$\epsilon_I > 0 \qquad \text{for} \quad q_I^2\pi_I \geq 1\,.$$

On the other hand, positivity of $\epsilon_I = \epsilon_I(\delta_I)$ is given if and only if

$$\delta_I > 0 \qquad \text{for} \quad q_I^2\pi_I \leq 1\,,$$

$$0 < \delta_I < \frac{-q_I(1+q_I)}{1-q_I^2\pi_I} \qquad \text{for} \quad q_I^2\pi_I > 1\,.$$

Because of (3.16) and (3.17), positivity of $\rho_I = \rho_I(\tau_I)$ and of $\sigma_I = \sigma_I(\tau_I)$ is given if and only if

$$\tau_I > \frac{(\epsilon_I + q_I)(\epsilon_I + \delta_I)}{3(1+q_I)\epsilon_I}\,.$$

On the other hand, because of (3.15) and (3.16), positivity of $\rho_I = \rho_I(\sigma_I)$ and of $\tau_I = \tau_I(\sigma_I)$ is given if and only if

$$\sigma_I > \frac{\delta_I + \epsilon_I q_I}{3(1+q_I)}\,,$$

and because of (3.15) and (3.17), positivity of $\sigma_I = \sigma_I(\rho_I)$ and of $\tau_I = \tau_I(\rho_I)$ is given if and only if

$$\rho_I > \frac{(1 + \delta_I q_I \pi_I)(\epsilon_I + \delta_I)}{3(1+q_I)\delta_I \pi_I}\,.$$

The minimum norm characterization of interval weighted $\tau$-splines works for nonnegative tension parameter $\nu_{I,1}$ and $\nu_{I,2}$ and $p_I > 0$ only. As for interval weighted $\nu$-splines we can extend the theory by requiring *positivity of the shape parameters* of the B-spline-Bézier representation to ensure the convex hull and the variation diminishing properties.

Assume that $p_I > 0$, i.e., $\pi_I > 0$, then, because of (3.12), for $\epsilon_I > 0$ and $\delta_I > 0$ the tension parameters $\nu_{I,2}$ have to be within the range

$$\mu_{I,2} > \max\left\{-(1 + 2q_I\pi_I + q_I^2\pi_I)\frac{3}{\Delta_I}\,,\; -\frac{(1 + 2q_I + q_I^2\pi_I)}{q_I}\frac{3}{\Delta_I}\right\}\,.$$

Thus, not only nonnegative but also certain negative $\nu_{I,2}$-values are allowed.

Because of (3.13), for $\rho_I > 0$, $\sigma_I > 0$ and $\tau_I > 0$ the tension parameters $\nu_{I,1}$ have to be within the ranges

$$\mu_{I,1} \;<\; \frac{24}{\triangle_I^3}\left[1 + 3q_I + 3q_I^2\pi_I + q_I^3\pi_I + q_I\triangle_I\mu_{I,2}\right]\Pi_I\,,$$

$$\mu_{I,1} \;>\; -\frac{24}{\triangle_I^3}\frac{1 + 4q_I\pi_I + 6q_I^2\pi_I + 4q_I^3\pi_I + q_I^4\pi_I^2 + 2(1 + q_I^3\pi_I)\frac{\triangle_I}{3}\mu_{I,2}}{2(1 + q_I\pi_I) + \frac{\triangle_I}{3}\mu_{I,2}}\,,$$

where $\Pi_I = \max\{1,\pi_I\}$. That means first the $\nu_{I,2}$ have to be chosen such that $\epsilon_I > 0$ and $\delta_I > 0$ are fulfilled, and then the $\nu_{I,1}$ can be chosen within the ranges given above.

The ten auxiliary parameters of the B-spline construction also have to be positive to ensure properties like the ones mentioned above. Therefore, some more restrictions (cf. [1]) have to be fulfilled.

## References

[1] M. Eck and D. Lasser, *B-Spline-Bézier Representation of Geometric Spline Curves. Quartics and Quintics*, Preprint 1254, Fachbereich Mathematik, Technische Hochschule Darmstadt, Darmstadt, 1989.

[2] T. A. Foley, *Local control of interval tension using weighted splines*, Comput. Aided Geom. Des., 3 (1986), pp. 281–294.

[3] ——, *Interpolation with interval and point tension controls using cubic weighted ν-splines*, ACM Trans. Math. Software, 13 (1987), pp. 68–96.

[4] H. Hagen, *Geometric spline curves*, Comput, Aided Geom. Des., 2 (1985), pp. 223–227, Also Technical Report TR-85-011, Arizona State University Department of Computer Science, Tempe, 1985.

[5] J. Hoschek and D. Lasser, *Grundlagen der Geometrischen Datenverarbeitung*, Teubner, 1989.

[6] D. Lasser, *B-Spline-Bézier Representation of Tau-Splines*, Technical Report NPS-53-88-006, Naval Postgraduate School, Monterey, 1988.

[7] G. M. Nielson, *Some piecewise polynomial alternatives to splines under tension*, in Computer Aided Geometric Design, R. E. Barnhill and R. F. Riesenfeld, eds., Academic Press, Orlando, 1974, pp. 209–235.

[8] D. A. Neuser, *Curve and surface interpolation using weighted quintic tau-splines*, in Curve and Surface Design, H. Hagen, ed., Society for Industrial and Applied Mathematics, Philadelphia, 1992, pp. 55–85.

[9] H. Pottmann, *Smooth curves under tension*, Comput. Aided Des., 22 (1990), pp. 241–245.

[10] K. Salkauskas, *$C^1$-splines for interpolation of rapidly varying data*, Rocky Mountain J. Math., 14 (1974), pp. 239–250.

# Curve and Surface Interpolation Using Quintic Weighted Tau-Splines

David A. Neuser

## 4.1. Introduction

Given the data points $(t_i, y_i)$ for $i = 1, \cdots, n$, the *natural cubic spline* is the $C^2$ piecewise cubic interpolating function that minimizes

$$\int_{t_1}^{t_n} \left[f''(t)\right]^2 dt$$

over all functions $f$ in the space $H^2 = \{f | f'' \in L^2[t_1, t_n]$ and $f'$ is absolutely continuous on $[t_1, t_n]\}$ that satisfy $f(t_i) = y_i$, $i = 1, \cdots, n$. In 1966, Schweikert [20] first proposed splines under tension as a solution to removing some unwanted undulations in the interpolating curve. The *spline under tension* is the interpolating function in $H^2[t_1, t_n]$ that minimizes

$$\int_{t_1}^{t_n} \left[f''(t)\right]^2 dt \quad + \quad s \int_{t_1}^{t_n} \left[f'(t)\right]^2 dt$$

and it involves both polynomial and exponential functions. As the values of the global parameter $s$ increase from 0 to $\infty$, the curve varies from the natural cubic spline to the polygonal line segments connecting the data points. In 1973, Pilcher [17],[18] and later in 1984, Barsky [4] constructed an exponential-based spline under tension as the interpolating function in $H^2[t_1, t_n]$ that minimizes

$$\int_{t_1}^{t_n} \left[f''(t)\right]^2 dt + \sum_{i=1}^{n-1} w_i \int_{t_i}^{t_{i+1}} \left[f'(t)\right]^2 dt$$

where $w_i \geq 0$, for $i = 1, \cdots, n-1$. This minimizing function is a piecewise combination of an exponential and linear function that belongs to $C^2$. As $w_i$ increases without bound, only the $i$th interval begins to be drawn to a straight line segment. However, this spline function is expensive to evaluate due to its exponential component. Nielson [15] found a piecewise polynomial alternative to splines in tension by finding the piecewise cubic interpolating function in $H^2[t_1, t_n]$ that minimizes

$$\int_{t_1}^{t_n} \left[f''(t)\right]^2 dt + \sum_{i=1}^{n} v_i \left[f'(t_i)\right]^2$$

where $v_i \geq 0$ for $i = 1, \cdots, n$. In addition, he found that even though the unique solution satisfying the interpolation and end conditions does not have continuous second-order derivatives, it does, in the parametric case, have continuous curvature. When applied to a parametric curve, these so-called $\nu$-*splines* allow the curve to be "tightened" at the $i$th interpolation point by increasing the value of $v_i$, thus creating a "corner" at that point. When $v_i = 0$ for $i = 1, \cdots, n$, the $C^1$ $\nu$-spline reduces to the $C^2$ natural cubic spline. In 1984, Salkauskas [19] generalized the natural cubic spline by finding the interpolating function which minimizes

$$\sum_{i=1}^{n-1} w_i \int_{t_i}^{t_{i+1}} \left[f''(t)\right]^2 dt$$

over all $f \in H^2$ and $w_i > 0$ for $i = 1, \cdots, n-1$. This $C^1$ *weighted spline* provides for "tightening" the curve across an interval and reduces to the natural cubic spline when all the interval weights are equal. In 1987, Foley [9] generalized both the weighted spline and the $\nu$-spline by finding the interpolating function in $H^2$ that minimizes

$$\sum_{i=1}^{n-1} w_i \int_{t_i}^{t_{i+1}} \left[f''(t)\right]^2 dt + \sum_{i=1}^{n} v_i \left[f'(t_i)\right]^2$$

where $w_i > 0$ for $i = 1, \cdots, n-1$ and $v_i \geq 0$ for $i = 1, \cdots, n$. The resulting *weighted $\nu$-spline* is a $C^1$ piecewise cubic function which provides shape control parameters at the control points and across intervals between consecutive control points. With the proper choices for the values of the $w_i$s and the $v_i$s, the weighted $\nu$-spline reduces to either the weighted spline, the $\nu$-spline, or the natural cubic spline. Hagen [13] has generalized the $\nu$-splines to geometric spline curves and gives the computational equations for quintic spline curves with both curvature and torsion continuity. In the quintic case, the so-called $\tau$-*spline* is the interpolating function that minimizes

$$\int_{t_1}^{t_n} \left[f'''(t)\right]^2 dt + \sum_{i=1}^{n} v_i^1 \left[f'(t_i)\right]^2 + \sum_{i=1}^{n} v_i^2 \left[f''(t_i)\right]^2$$

where $v_i^1 \geq 0$ and $v_i^2 \geq 0$ for $i = 1, \cdots, n$.*

---

* Throughout this paper, what appear to be exponents on the variables $v$ and $w$ are to be interpreted as superscripts.

An objective of this paper is to generalize Hagen's quintic $\tau$-spline to include an interval weight factor on the third derivative.

## 4.2. The Piecewise Quintic Hermite Interpolant.

Suppose we are given the real values $y_i, y_i'$, and $y_i''$ for $i = 1, \cdots, n$, then there exists a unique $C^2$ piecewise quintic polynomial $f(t)$ defined on $[t_1, t_n]$ such that $f(t_i) = y_i$, $f'(t_i) = y_i'$, and $f''(t_i) = y_i''$. One representation for $f(t)$ comes from the well-known quintic Hermite blending functions

$$
\begin{aligned}
H_1(t) &= -6t^5 + 15t^4 - 10t^3 + 1, & H_4(t) &= -3t^5 + 7t^4 - 4t^3, \\
H_2(t) &= 6t^5 - 15t^4 + 10t^3 & H_5(t) &= \tfrac{1}{2}(-t^5 + 3t^4 - 3t^3 + t^2), \\
H_3(t) &= -3t^5 + 8t^4 - 6t^3 + t, & H_6(t) &= \tfrac{1}{2}(t^5 - 2t^4 + t^3),
\end{aligned}
$$

defined on $[0,1]$. By setting

$$
\widehat{H}_k(t) = H_k\left(\frac{t - t_i}{t_{i+1} - t_i}\right), \qquad k = 1, \cdots, 6,
$$

the blending functions can be defined over the more arbitrary interval $[t_i, t_{i+1}]$. If we let $h_i = t_{i+1} - t_i$, then

$$
\begin{aligned}
\widehat{H}_1(t) &= (t_{i+1} - t)^3 \left[6(t - t_i)^2 + 3(t - t_i)h_i + h_i^2\right] h_i^{-5}, \\
\widehat{H}_2(t) &= (t - t_i)^3 \left[6(t_{i+1} - t)^2 + 3(t_{i+1} - t)h_i + h_i^2\right] h_i^{-5}, \\
\widehat{H}_3(t) &= (t_{i+1} - t)^3 (t - t_i) \left[3(t - t_i) + h_i\right] h_i^{-5}, \\
\widehat{H}_4(t) &= (t_{i+1} - t)(t - t_i)^3 \left[3(t - t_{i+1}) - h_i\right] h_i^{-5}, \\
\widehat{H}_5(t) &= \frac{1}{2}(t_{i+1} - t)^3 (t - t_i)^2 h_i^{-5}, \\
\widehat{H}_6(t) &= \frac{1}{2}(t_{i+1} - t)^2 (t - t_i)^3 h_i^{-5}
\end{aligned}
$$

for $t_i \leq t \leq t_{i+1}$.

The above transformation involves both a translation and a scaling in the variable $t$. Whereas the functional values are invariant under the above transformation, the derivatives are not invariant under scaling. As a result, the scaled first derivatives and the scaled second derivatives must be adjusted by factors of $h_i$ and $h_i^2$, respectively, to recover the original derivatives. Thus $f(t)$ on $[t_i, t_{i+1}]$ has the cardinal Hermite form

(4.1)
$$
f(t) = \widehat{H}_1(t)y_i + \widehat{H}_2(t)y_{i+1} + \widehat{H}_3(t)h_i y_i' + \widehat{H}_4(t)h_i y_{i+1}' + \widehat{H}_5(t)h_i^2 y_i'' + \widehat{H}_6(t)h_i^2 y_{i+1}''.
$$

Motivated by the relationships

$$
H_1(t) = H_2(1 - t), \qquad H_3(t) = -H_4(1 - t), \qquad \text{and} \quad H_5(t) = H_6(1 - t),
$$

we define the following functions:

$$P_i(t) = \begin{cases} (t - t_{i-1})^3 \left[6(t_i - t)^2 + 3(t_i - t)h_{i-1} + h_{i-1}^2\right] h_{i-1}^{-5}, & \text{if } t_{i-1} \leq t \leq t_i; \\ (t_{i+1} - t)^3 \left[6(t - t_i)^2 + 3(t - t_i)h_i + h_i^2\right] h_i^{-5}, & \text{if } t_i < t \leq t_{i+1}; \\ 0 & \text{elsewhere.} \end{cases}$$

$$D_i(t) = \begin{cases} (t_i - t)(t - t_{i-1})^3 \left[3(t - t_i) - h_{i-1}\right] h_{i-1}^{-4}, & \text{if } t_{i-1} \leq t \leq t_i; \\ (t_{i+1} - t)^3 (t - t_i) \left[3(t - t_i) + h_i\right] h_i^{-4}, & \text{if } t_i < t \leq t_{i+1}; \\ 0 & \text{elsewhere.} \end{cases}$$

$$Q_i(t) = \begin{cases} \frac{1}{2}(t_i - t)^2 (t - t_{i-1})^3 h_{i-1}^{-3}, & \text{if } t_{i-1} \leq t \leq t_i; \\ \frac{1}{2}(t_{i+1} - t)^3 (t - t_i)^2 h_i^{-3}, & \text{if } t_i < t \leq t_{i+1}; \\ 0 & \text{elsewhere.} \end{cases}$$

The interpolant defined on $[t_1, t_n]$ now has the form

$$f(t) = \sum_{i=1}^{n} \left[y_i P_i(t) + y_i' D_i(t) + y_i'' Q_i(t)\right].$$

Since these modified quintic Hermite basis functions and their first and second derivatives exhibit the Kronecker-$\delta$ property

$$\begin{aligned} P_i(t_j) &= 1, & D_i(t_j) &= 0, & Q_i(t_j) &= 0, \\ P_i'(t_j) &= 0, & D_i'(t_j) &= 1, & Q_i'(t_j) &= 0, \\ P_i''(t_j) &= 0, & D_i''(t_j) &= 0, & Q_i''(t_j) &= 1, \end{aligned}$$

for $i = j$ and are all zero for $i \neq j$, it is easy to verify that $f(t_i) = y_i$, $f'(t_i) = y_i'$, and $f''(t_i) = y_i''$ for $i = 1, \cdots, n$. Furthermore, since $P_i(t)$, $D_i(t)$, and $Q_i(t)$ are zero outside of $[t_{i-1}, t_{i+1}]$, $i = 2, \cdots, n - 1$ , the $C^2$ interpolant $f(t)$ simplifies to

$$f(t) = y_i P_i(t) + y_i' D_i(t) + y_i'' Q_i(t) + y_{i+1} P_{i+1}(t) + y_{i+1}' D_{i+1}(t) + y_{i+1}'' Q_{i+1}(t)$$

for $t \in [t_i, t_{i+1}]$.

### 4.3. The Weighted Tau-Spline

Suppose we are given the sequence of points $\{(t_i, y_i)\}_{i=1}^{i=n}$ where $t_1 < \cdots < t_n$, and interval weights $w_i > 0$ for $i = 1, \cdots, n - 1$, and point tension values $v_i^1 \geq 0$ and $v_i^2 \geq 0$ for $i = 1, \cdots, n$. The *weighted $\tau$-spline* interpolant is the $C^2$ piecewise quintic interpolant $S(t)$ that minimizes

$$V(f) = \sum_{i=1}^{n-1} \int_{t_i}^{t_{i+1}} w_i [f'''(t)]^2 \, dt + \sum_{i=1}^{n} v_i^1 [f'(t_i)]^2 + \sum_{i=1}^{n} v_i^2 [f''(t_i)]^2$$

over all functions $f(t)$ in $H^3[t_1, t_n]$, subject to the interpolating conditions $f(t_i) = y_i$ for $i = 1, \cdots, n$, and some end conditions which are discussed later in this section. The set $H^3[t_1, t_n] = \{f | f''' \in L^2[t_1, t_n]$ and $f''$ is absolutely continuous on $[t_1, t_n]\}$ contains all the $C^2$ piecewise quintics. To find $S(t)$, one approach is to set up and solve the system of normal equations generated by

$$\frac{\partial V(f)}{\partial y_i'} = 0 \quad \text{and} \quad \frac{\partial V(f)}{\partial y_i''} = 0$$

for $i = 2, \cdots, n-1$. If we let

$$f_i(t) =$$
$$y_i P_i(t) + y_i' D_i(t) + y_i'' Q_i(t) + y_{i+1} P_{i+1}(t) + y_{i+1}' D_{i+1}(t) + y_{i+1}'' Q_{i+1}(t)$$

be the piecewise quintic interpolant defined on $[t_i, t_{i+1}]$, then

$$f(t) = \begin{cases} f_1(t), & \text{if } t \in [t_1, t_2] \\ f_2(t), & \text{if } t \in [t_2, t_3] \\ \vdots \\ f_{n-1}(t), & \text{if } t \in [t_{n-1}, t_n]. \end{cases}$$

Since

$$f_i^{(j)}(t) =$$
$$y_i P_i^{(j)}(t) + y_i' D_i^{(j)}(t) + y_i'' Q_i^{(j)}(t) + y_{i+1} P_{i+1}^{(j)}(t) + y_{i+1}' D_{i+1}^{(j)}(t) + y_{i+1}'' Q_{i+1}^{(j)}(t)$$

for $t \in (t_i, t_{i+1}), i = 1, \cdots, n-1$, and $f_i'(t_i) = y_i'$ and $f_i''(t_i) = y_i''$ for $i = 1, \cdots, n$, it follows that

$$V(f) = \sum_{i=1}^{n-1} w_i \int_{t_i}^{t_{i+1}} \left[ f_i'''(t) \right]^2 dt + \sum_{i=1}^{n} v_i^1 (y_i')^2 + \sum_{i=1}^{n} v_i^2 (y_i'')^2.$$

Thus

$$\frac{\partial V(f)}{\partial y_k'} = \frac{\partial}{\partial y_k'} \left\{ w_{k-1} \int_{t_{k-1}}^{t_k} \left[ f_{k-1}'''(t) \right]^2 dt + w_k \int_{t_k}^{t_{k+1}} \left[ f_k'''(t) \right]^2 dt + v_k^1 (y_k')^2 \right\}$$

$$= 2w_{k-1} \int_{t_{k-1}}^{t_k} \left[ f_{k-1}'''(t) \right] D_k'''(t) dt + 2w_k \int_{t_k}^{t_{k+1}} \left[ f_k'''(t) \right] D_k'''(t) dt + 2v_k^1 y_k'$$

$$= 0$$

Using MACSYMA, an expert system that does symbolic mathematical manipulation, to integrate and simplify the 12 products of third derivatives, the following equations emerge:

(4.2)
$$168 w_{k-1} h_{k-1}^{-3} y_{k-1}' + 24 w_{k-1} h_{k-1}^{-2} y_{k-1}'' + (192 w_{k-1} h_{k-1}^{-3} + 192 w_k h_k^{-3} + v_k^1) y_k'$$
$$+ 36(-w_{k-1} h_{k-1}^{-2} + w_k h_k^{-2}) y_k'' + 168 w_k h_k^{-3} y_{k+1}' - 24 w_k h_k^{-2} y_{k+1}''$$
$$= 360 \left[ w_{k-1} h_{k-1}^{-4} (y_k - y_{k-1}) + w_k h_k^{-4} (y_{k+1} - y_k) \right]$$

where $k = 2, \cdots, n-1$. Similarly,

$$
\frac{\partial V(f)}{\partial y_k''} = \frac{\partial}{\partial y_k''} \left\{ w_{k-1} \int_{t_{k-1}}^{t_k} \left[ f_{k-1}'''(t) \right]^2 dt + w_k \int_{t_k}^{t_{k+1}} \left[ f_k'''(t) \right]^2 dt + v_k^2 (y_k'')^2 \right\}
$$

$$
= 2w_{k-1} \int_{t_{k-1}}^{t_k} \left[ f_{k-1}'''(t) \right] Q_k'''(t) dt + 2w_k \int_{t_k}^{t_{k+1}} \left[ f_k'''(t) \right] Q_k'''(t) dt + 2v_k^2 y_k''
$$

$$
= 0
$$

yields the additional equations

$$
\begin{aligned}
&- 24 w_{k-1} h_{k-1}^{-2} y_{k-1}' - 3 w_{k-1} h_{k-1}^{-1} y_{k-1}'' + 36(-w_{k-1} h_{k-1}^{-2} + w_k h_k^{-2}) y_k' \\
(4.3) \quad &+ (9 w_{k-1} h_{k-1}^{-1} + 9 w_k h_k^{-1} + v_k^2) y_k'' + 24 w_k h_k^{-2} y_{k+1}' - 3 w_k h_k^{-1} y_{k+1}'' \\
&= 60 \left[ w_{k-1} h_{k-1}^{-3}(y_{k-1} - y_k) + w_k h_k^{-3}(y_{k+1} - y_k) \right]
\end{aligned}
$$

where $k = 2, \cdots, n-1$.

This gives us $2n-4$ equations in the $2n$ unknowns $y_i'$ and $y_i''$, for $i = 1 \cdots n$. The remaining four equations come from the given end conditions. For Type I conditions, commonly referred to as *derivative* or "clamped" end conditions, the four additional equations are

$$
\begin{aligned}
y_1' &= m_1, & y_n' &= m_n, \\
y_1'' &= M_1, & y_n'' &= M_n.
\end{aligned}
$$

The Type II conditions, or *natural* end conditions, are given by

$$
\begin{aligned}
w_1 y'''(t_1^+) - v_1^2 y_1'' &= 0, \\
w_{n-1} y'''(t_n^-) + v_n^2 y_n'' &= 0, \\
w_1 y^{(4)}(t_1^+) + v_1^1 y_1' &= 0, \\
w_{n-1} y^{(4)}(t_n^-) - v_n^1 y_n' &= 0.
\end{aligned}
$$

Using the fact that

$$
\begin{aligned}
f^{(4)}(t_1^+) &= 12 h_1^{-2} \left[ 30 h_1^{-2}(y_1 - y_2) + 2 h_1^{-1}(8 y_1' + 7 y_2') + (3 y_1'' - 2 y_2'') \right], \\
f'''(t_1^+) &= 3 h_1^{-1} \left[ 20 h_1^{-2}(y_2 - y_1) - 4 h_1^{-1}(2 y_2' + 3 y_1') + (y_2'' - 3 y_1'') \right], \\
f^{(4)}(t_n^-) &= 12 h_{n-1}^{-2} \left[ 30 h_{n-1}^{-2}(y_n - y_{n-1}) - 2 h_{n-1}^{-1}(8 y_n' + 7 y_{n-1}') + (3 y_n'' - 2 y_{n-1}'') \right], \\
f'''(t_n^-) &= 3 h_{n-1}^{-1} \left[ 20 h_{n-1}^{-2}(y_n - y_{n-1}) - 4 h_{n-1}^{-1}(3 y_n' + 2 y_{n-1}') + (3 y_n'' - y_{n-1}'') \right],
\end{aligned}
$$

the Type II end conditions yield the following four equations

$$
(192 w_1 h_1^{-3} + v_1^1) y_1' + 36 w_1 h_1^{-2} y_1'' + 168 w_1 h_1^{-3} y_2' - 24 w_1 h_1^{-2} y_2''
$$
$$
= 360 w_1 h_1^{-4}(y_2 - y_1),
$$

$$
36 w_1 h_1^{-2} y_1' + (9 w_1 h_1^{-1} + v_1^2) y_1'' + 24 w_1 h_1^{-2} y_2' - 3 w_1 h_1^{-1} y_2''
$$
$$
= 60 w_1 h_1^{-3}(y_2 - y_1),
$$

$$
(4.4) \quad 168 w_{n-1} h_{n-1}^{-3} y_{n-1}' + 24 w_{n-1} h_{n-1}^{-2} y_{n-1}'' + (192 w_{n-1} h_{n-1}^{-3} + v_n^1) y_n'
$$
$$
- 36 w_{n-1} h_{n-1}^{-2} y_n'' = 360 w_{n-1} h_{n-1}^{-4}(y_n - y_{n-1}),
$$

$$
-24 w_{n-1} h_{n-1}^{-2} y_{n-1}' - 3 w_{n-1} h_{n-1}^{-1} y_{n-1}'' - 36 w_{n-1} h_{n-1}^{-2} y_n'
$$
$$
+ (9 w_{n-1} h_{n-1}^{-1} + v_n^2) y_n'' = 60 w_{n-1} h_{n-1}^{-3}(y_{n-1} - y_n).
$$

In general, if $S(t)$ is a natural spline of degree $2k - 1$ defined on $[t_1, t_n]$, then $S(t)$ is of degree $k - 1$ or less on $(-\infty, t_1]$ and $[t_n, \infty)$. If $f(t)$ is a natural fifth degree spline, then it follows that $f(t)$ is at most a quadratic prior to the point $(t_1, y_n)$ and at most a quadratic following the point $(t_n, y_n)$. This implies that $y'''(t_1^-)$, $y'''(t_n^+)$, $y^{(4)}(t_1^-)$, and $y^{(4)}(t_n^+)$ are all zero.

The four equations for natural end conditions are equivalent to requiring that

$$\frac{\partial V(f)}{\partial y_k'} = 0 \quad \text{and} \quad \frac{\partial V(f)}{\partial y_k''} = 0$$

for $k = 1$ and $k = n$, which is equivalent to minimizing $V(f)$ without subjecting $f(t)$ to any end conditions.

The Type III conditions, or *periodic* end conditions, are given by

$$y_1' = y_n',$$
$$y_1'' = y_n'',$$
$$w_1 y'''(t_1^+) - w_{n-1} y'''(t_n^-) = v_1^2 y_1'' + v_n^2 y_n'',$$
$$w_{n-1} y^{(4)}(t_n^-) - w_1 y^{(4)}(t_1^+) = v_1^1 y_1' + v_n^1 y_n'$$

and yield the four additional equations

$$(192 w_1 h_1^{-3} + 192 w_{n-1} h_{n-1}^{-3} + v_1^1 + v_{n-1}^1) y_1' + (36 w_1 h_1^{-2} - 36 w_{n-1} h_{n-1}^{-2}) y_1''$$
$$+ 168 w_1 h_1^{-3} y_2' - 24 w_1 h_1^{-2} y_2'' + 168 w_{n-1} h_{n-1}^{-3} y_{n-1}' - 24 w_{n-1} h_{n-1}^{-2} y_{n-1}''$$
$$= 360 w_1 h_1^{-4}(y_2 - y_1) + 360 w_{n-1} h_{n-1}^{-4}(y_n - y_{n-1}),$$
$$(36 w_1 h_1^{-2} - 36 w_{n-1} h_{n-1}^{-2}) y_1' + (9 w_1 h_1^{-1} + 9 w_{n-1} h_{n-1}^{-1} + v_1^2 + v_{n-1}^2) y_1''$$
$$+ 24 w_1 h_1^{-2} y_2' - 3 w_1 h_1^{-1} y_2'' - 24 w_{n-1} h_{n-1}^{-2} y_2' - 3 w_{n-1} h_{n-1}^{-1} y_2''$$
$$= 60 w_1 h_1^{-3}(y_2 - y_1) - 60 w_{n-1} h_{n-1}^{-3}(y_n - y_{n-1}),$$
$$y_1' = y_n', \quad \text{and} \quad y_1'' = y_{n-1}''.$$

If we let $m = 2n$, then the resulting system of $m$ equations in $m$ unknowns has the following matrix form:

$$
\begin{bmatrix}
d_1 & a_1 & e_1 & g_1 \\
a_1 & d_2 & a_2 & e_2 \\
e_1 & a_2 & d_3 & a_3 & e_3 & g_3 \\
g_1 & e_2 & a_3 & d_4 & a_4 & e_4 \\
& & e_3 & a_4 & d_5 & a_5 & e_5 & g_5 \\
& & g_3 & e_4 & a_5 & d_6 & a_6 & e_6 \\
& & & & & & \ddots \\
& & & & e_{m-7} & a_{m-6} & d_{m-5} & a_{m-5} & e_{m-5} & g_{m-5} \\
& & & & g_{m-7} & e_{m-6} & a_{m-5} & d_{m-4} & a_{m-4} & e_{m-4} \\
& & & & & & e_{m-5} & a_{m-4} & d_{m-3} & a_{m-3} & e_{m-3} & g_{m-3} \\
& & & & & & g_{m-5} & e_{m-4} & a_{m-3} & d_{m-2} & a_{m-2} & e_{m-2} \\
& & & & & & & & e_{m-3} & a_{m-2} & d_{m-1} & a_{m-1} \\
& & & & & & & & g_{m-3} & e_{m-2} & a_{m-1} & d_m
\end{bmatrix}
$$

$$(4.5) \qquad \begin{bmatrix} y_1' \\ y_1'' \\ y_2' \\ y_2'' \\ \\ \cdot \\ \cdot \\ \cdot \\ \\ \\ y_{n-1}' \\ y_{n-1}'' \\ y_n' \\ y_n'' \end{bmatrix} = \begin{bmatrix} b_1 \\ b_2 \\ b_3 \\ b_4 \\ \\ \cdot \\ \cdot \\ \cdot \\ \\ \\ b_{m-3} \\ b_{m-2} \\ b_{m-1} \\ b_m \end{bmatrix},$$

where if we define $w_0 = w_n = y_0 = y_{n+1} = 0$ and $h_0 = h_n = 1$, then

$$d_{2i-1} = 192(w_{i-1}h_{i-1}^{-3} + w_i h_i^{-3}) + v_i^1, \quad i = 1, \cdots, n,$$
$$d_{2i} = 9(w_{i-1}h_{i-1}^{-1} + w_i h_i^{-1}) + v_i^2, \quad i = 1, \cdots, n,$$
$$a_{2i-1} = 36(-w_{i-1}h_{i-1}^{-2} + w_i h_i^{-2}), \quad i = 1, \cdots, n,$$
$$a_{2i} = 24 w_i h_i^{-2}, \quad i = 1, \cdots, n-1,$$
$$e_{2i-1} = 168 w_i h_i^{-3}, \quad i = 1, \cdots, n-1,$$
$$e_{2i} = -3 w_i h_i^{-1}, \quad i = 1, \cdots, n-1,$$
$$g_{2i-1} = -24 w_i h_i^{-2}, \quad i = 1, \cdots, n-1,$$
$$b_{2i-1} = 360 \left[ w_{i-1}h_{i-1}^{-4}(y_i - y_{i-1}) + w_i h_i^{-4}(y_{i+1} - y_i) \right], \quad i = 1, \cdots, n,$$
$$b_{2i} = 60 \left[ -w_{i-1}h_{i-1}^{-3}(y_i - y_{i-1}) + w_i h_i^{-3}(y_{i+1} - y_i) \right], \quad i = 1, \cdots, n.$$

Even though the system of equations in (4.5) is not diagonally dominant, it appears to be stable. A general method for identifying certain ranges (positive and negative) of weighted $\tau$-spline parameters which generate nonsingular systems of equations is given in [14]. Since this system is banded, it can be solved in linear time.

For derivative end conditions, i.e., given $y_1' = S'(t_1)$, $y_1'' = S''(t_1)$, $y_n' = S'(t_n)$, and $y_n'' = S''(t_n)$, we need to solve the following system of $2n - 4$ equations for the $2n - 4$ unknowns $y_i'$ and $y_i''$, where $i = 2, \cdots, n-1$:

$$
\begin{bmatrix}
d_3 & a_3 & e_3 & g_3 \\
a_3 & d_4 & a_4 & e_4 \\
e_3 & a_4 & d_5 & a_5 & e_5 & g_5 \\
g_3 & e_4 & a_5 & d_6 & a_6 & e_6 \\
 & & e_5 & a_6 & d_7 & a_7 & e_7 & g_7 \\
 & & g_5 & e_6 & a_7 & d_8 & a_8 & e_8 \\
 & & & & & & \ddots \\
 & & & & e_{m-9} & a_{m-8} & d_{m-7} & a_{m-7} & e_{m-7} & g_{m-7} \\
 & & & & g_{m-9} & e_{m-8} & a_{m-7} & d_{m-6} & a_{m-6} & e_{m-6} \\
 & & & & & & e_{m-7} & a_{m-6} & d_{m-5} & a_{m-5} & e_{m-5} & g_{m-5} \\
 & & & & & & g_{m-7} & e_{m-6} & a_{m-5} & d_{m-4} & a_{m-4} & e_{m-4} \\
 & & & & & & & & e_{m-5} & a_{m-4} & d_{m-3} & a_{m-3} \\
 & & & & & & & & g_{m-5} & e_{m-4} & a_{m-3} & d_{m-2}
\end{bmatrix}
$$

$$
\begin{bmatrix}
y_2' \\
y_2'' \\
y_3' \\
y_3'' \\
\vdots \\
y_{n-2}' \\
y_{n-2}'' \\
y_{n-1}' \\
y_{n-1}''
\end{bmatrix}
=
\begin{bmatrix}
b_3 - e_1 S'(t_1) - a_2 S''(t_1) \\
b_4 - g_1 S'(t_1) - e_2 S''(t_1) \\
b_3 \\
b_4 \\
\vdots \\
b_{m-5} \\
b_{m-4} \\
b_{m-3} - e_{m-3} S'(t_n) - g_{m-3} S''(t_n) \\
b_{m-2} - a_{m-2} S'(t_n) - e_{m-2} S''(t_n)
\end{bmatrix}
$$

For periodic end conditions, i.e., given that $y_1' = S'(t_1) = S'(t_n)$ and $y_1'' = S''(t_1) = S''(t_n)$, we need to solve the following system of $2n-2$ equations for the $2n-2$ unknowns $y_i'$ and $y_i''$, where $i = 1, \cdots, n-1$:

$$
\begin{bmatrix}
d_1+d_{m-1} & a_1+a_{m-1} & e_1 & g_1 & & & & & e_{m-3} & a_{m-2} \\
a_1+a_{m-1} & d_2+d_m & a_2 & e_2 & & & & & g_{m-3} & e_{m-2} \\
e_1 & a_2 & d_3 & a_3 & e_3 & g_3 & & & & \\
g_1 & e_2 & a_3 & d_4 & a_4 & e_4 & & & & \\
 & & & & \cdot & & & & & \\
 & & & & & \cdot & & & & \\
 & & & & & & \cdot & & & \\
 & & & e_{m-7} & a_{m-6} & d_{m-5} & a_{m-5} & e_{m-5} & g_{m-5} \\
 & & & g_{m-7} & e_{m-6} & a_{m-5} & d_{m-4} & a_{m-4} & e_{m-4} \\
e_{m-3} & g_{m-3} & & & & e_{m-5} & a_{m-4} & d_{m-3} & a_{m-3} \\
a_{m-2} & e_{m-2} & & & & g_{m-5} & e_{m-4} & a_{m-3} & d_{m-2}
\end{bmatrix}
$$

$$
\begin{bmatrix}
y_1' \\ y_1'' \\ y_2' \\ y_2'' \\ \vdots \\ \\ y_{n-2}' \\ y_{n-2}'' \\ y_{n-1}' \\ y_{n-1}''
\end{bmatrix}
=
\begin{bmatrix}
b_1+b_{m-1} \\ b_2+b_m \\ b_3 \\ b_4 \\ \vdots \\ \\ b_{m-5} \\ b_{m-4} \\ b_{m-3} \\ b_{m-2}
\end{bmatrix}
$$

Algorithms in the form of Pascal procedures for solving each of these systems of equations are given in [14]. The algorithms use customized Gaussian elimination techniques and the procedures run in linear time.

The following two theorems establish the relationships for the jump discontinuities in the third and fourth derivatives of the weighted $\tau$-spline.

THEOREM 4.3.1. *If $S(t)$ is any $C^2$ weighted $\tau$-spline that minimizes $V(f)$, then*

$$
w_k S'''(t_k^+) - w_{k-1} S'''(t_k^-) = v_k^2 S''(t_k)
$$

*for $k = 2, \cdots, n-1$.*

*Proof.* If $f(t)$ is any $C^2$ piecewise quintic, then

$$
f(t) = \sum_{i=1}^{n} [y_i P_i(t) + y_i' D_i(t) + y_i'' Q_i'''(t)]
$$

and

$$f'''(t_k^+) = 3h_k^{-1}\left[20h_k^{-2}(y_{k+1} - y_k) - 4h_k^{-1}(2y_{k+1}' + 3y_k') + (y_{k+1}'' - 3y_k'')\right]$$

and

$$f'''(t_k^-) = 3h_{k-1}^{-1}\left[20h_{k-1}^{-2}(y_k - y_{k-1}) - 4h_{k-1}^{-1}(3y_k' + 2y_{k-1}') + (3y_k'' - y_{k-1}'')\right]$$

for $k = 2, \cdots, n - 1$.

Thus, substituting in from (4.3), it follows that

$$\begin{aligned}
& w_k S'''(t_k^+) - w_{k-1} S'''(t_k^-) \\
&= 60\left\{w_{k-1}h_{k-1}^{-3}\left[S(t_{k-1}) - S(t_k)\right] + w_k h_k^{-3}\left[S(t_{k+1}) - S(t_k)\right]\right\} \\
&\quad + 24w_{k-1}h_{k-1}^{-2}S'(t_{k-1}) + 3w_{k-1}h_{k-1}^{-1}S''(t_{k-1}) \\
&\quad - 36(-w_{k-1}h_{k-1}^{-2} + w_k h_k^{-2})S'(t_k) - (9w_{k-1}h_{k-1}^{-1} + 9w_k h_k^{-1})S''(t_k) \\
&\quad - 24w_k h_k^{-2}S'(t_{k+1}) + 3w_k h_k^{-1}S''(t_{k+1}) \\
&= v_k^2 S''(t_k)
\end{aligned}$$

for $k = 2, \cdots, n - 1$.

If $S(t)$ is any $C^2$ weighted $\tau$-spline and if for some value of $k$, $k = 2, \cdots, n-1$, we have that $w_{k-1} = w_k = c$, a positive constant, and $v_k^2 = 0$, then it follows from Theorem 4.3.1 that $S'''(t_k^+) = S'''(t_k^-)$ . Thus $S(t)$ is $C^3$ on $(t_{k-1}, t_{k+1})$.

THEOREM 4.3.2. *If $S(t)$ is any $C^2$ weighted $\tau$-spline that minimizes $V(f)$, then*

$$w_{k-1}S^{(4)}(t_k^-) - w_k S^{(4)}(t_k^+) = v_k^1 S'(t_k)$$

*for $k = 2, \cdots, n - 1$.*

*Proof.* If $f(t)$ is any $C^2$ piecewise quintic, then

$$f(t) = \sum_{i=1}^{n}\left[y_i P_i(t) + y_i' D_i(t) + y_i'' Q_i(t)\right]$$

and

$$f^{(4)}(t_k^+) = 12h_k^{-2}\left[30h_k^{-2}(y_k - y_{k+1}) + 2h_k^{-1}(8y_k' + 7y_{k+1}') + 3y_k'' - 2y_{k+1}''\right]$$

and

$$f^{(4)}(t_k^-) = 12h_{k-1}^{-2}\left[30h_{k-1}^{-2}(y_k - y_{k-1}) - 2h_{k-1}^{-1}(8y_k' + 7y_{k-1}') + 3y_k'' - 2y_{k-1}''\right]$$

for $k = 2, \cdots, n - 1$.

Thus, substituting in from (4.3), it follows that

$$\begin{aligned}
& w_{k-1}S^{(4)}(t_k^-) - w_k S^{(4)}(t_k^+) \\
&= 360\left\{w_{k-1}h_{k-1}^{-4}\left[S(t_k) - S(t_{k-1})\right] + w_k h_k^{-4}\left[S(t_{k+1}) - S(t_k)\right]\right\} \\
&\quad - 168w_{k-1}h_{k-1}^{-3}S'(t_{k-1}) - 24w_{k-1}h_{k-1}^{-2}S''(t_{k-1}) \\
&\quad - 192(w_{k-1}h_{k-1}^{-3} + w_k h_k^{-3})S'(t_k) - 36(-w_{k-1}h_{k-1}^{-2} + w_k h_k^{-2})S''(t_k) \\
&\quad - 168w_k h_k^{-3}S'(t_{k+1}) + 24w_k h_k^{-2}S''(t_{k+1}) \\
&= v_k^1 S'(t_k)
\end{aligned}$$

for $k = 2, \cdots, n - 1$.

If $S(t)$ is any $C^2$ weighted $\tau$-spline and if for some value of $k$, $k = 2, \cdots, n - 1$, we have that $w_{k-1} = w_k = c$, a positive constant, and $v_k^2 = 0$, and $v_k^1 = 0$, then it follows from Theorems 4.3.1 and 4.3.2 that $S'''(t_k^+) = S'''(t_k^-)$ and $S^{(4)}(t_k^+) = S^{(4)}(t_k^-)$. Since these one-sided derivatives match, we have that $S'''(t)$ and $S^{(4)}(t)$ are continuous at $t_k$. Also note that if $v_k^2 = 0$ and $w_{k-1} = w_k$, then $S(t)$ is $C^3$ on $(t_{k-1}, t_{k+1})$.

The following theorem establishes the minimization property of the weighted $\tau$-spline over $H^3[t_1, t_n]$.

THEOREM 4.3.3. *If $g(t)$ is any function in $H^3[t_1, t_n]$ that satisfies $g(t_i) = y_i$ for $i = 1, \cdots, n$, and $S(t)$ is the natural weighted $\tau$-spline interpolant to $(t_i, y_i)$, $i = 1, \cdots, n$, then $V(S) \leq V(g)$. Furthermore, if $g(t)$ is any interpolant in $H^3[t_1, t_n]$ that satisfies Type I or Type III end conditions and $S(t)$ is the weighted $\tau$-spline interpolant satisfying the same end conditions, then $V(S) \leq V(g)$.*

*Proof.* We will first prove the analogue of the "first integral relation" for $C^4$ natural quintic splines that

$$V(g) = V(S) + V(g - S).$$

We proceed as in the proof of the minimum norm property in Ahlberg, Nilson, and Walsh [1]. Noting that $f^2 - R^2 - (f - R)^2 = 2R(f - R)$ , and then computing two levels of integration by parts over each subinterval $[t_i, t_{i+1}]$, it follows that

$$V(g) - V(S) - V(g - S)$$

$$= \sum_{i=1}^{n-1} \int_{t_i}^{t_{i+1}} w_i \left\{ [g'''(t)]^2 - [S'''(t)]^2 - [g'''(t) - S'''(t)]^2 \right\} dt$$

$$+ \sum_{i=1}^{n} v_i^1 \left\{ [g'(t_i)]^2 - [S'(t_i)]^2 - [g'(t_i) - S'(t_i)]^2 \right\}$$

$$+ \sum_{i=1}^{n} v_i^2 \left\{ [g''(t_i)]^2 - [S''(t_i)]^2 - [g''(t_i) - S''(t_i)]^2 \right\}$$

$$= 2 \sum_{i=1}^{n-1} w_i \int_{t_i}^{t_{i+1}} S'''(t)[g'''(t) - S'''(t)] dt$$

$$+ 2 \sum_{i=1}^{n} v_i^1 S'(t_i)[g'(t_i) - S'(t_i)] + 2 \sum_{i=1}^{n} v_i^2 S''(t_i)[g''(t_i) - S''(t_i)]$$

$$= 2 \sum_{i=1}^{n-1} w_i \left\{ \left[ S'''(t)[g''(t) - S''(t)] \right]_{t_i}^{t_{i+1}} - \left[ S^{(4)}(t)[g'(t) - S'(t)] \right]_{t_i}^{t_{i+1}} \right\}$$

$$+ 2 \sum_{i=1}^{n} v_i^1 S'(t_i)[g'(t_i) - S'(t_i)] + 2 \sum_{i=1}^{n} v_i^2 S''(t_i)[g''(t_i) - S''(t_i)]$$

$$= 2\sum_{i=1}^{n-1} w_i S'''(t_{i+1}^-)\big[g''(t_{i+1}) - S''(t_{i+1})\big] - 2\sum_{i=1}^{n-1} w_i S'''(t_i^+)\big[g''(t_i) - S''(t_i)\big]$$

$$- 2\sum_{i=1}^{n-1} w_i S^{(4)}(t_{i+1}^-)\big[g'(t_{i+1}) - S'(t_{i+1})\big] + 2\sum_{i=1}^{n-1} w_i S^{(4)}(t_i^+)\big[g'(t_i) - S'(t_i)\big]$$

$$+ 2\sum_{i=1}^{n} v_i^1 S'(t_i)[g'(t_i) - S'(t_i)] + 2\sum_{i=1}^{n} v_i^2 S''(t_i)[g''(t_i) - S''(t_i)]$$

$$= 2\sum_{i=2}^{n-1} \big[w_{i-1} S'''(t_i^-) - w_i S'''(t_i^+) + v_i^2 S''(t_i)\big]\big[g''(t_i) - S''(t_i)\big]$$

$$+ 2\sum_{i=2}^{n-1} \big[-w_{i-1} S^{(4)}(t_i^-) + w_i S^{(4)}(t_i^+) + v_i^1 S'(t_i)\big]\big[g'(t_i) - S'(t_i)\big]$$

$$- 2[w_1 S'''(t_1^+) - v_1^2 S''(t_1)][g''(t_1) - S''(t_1)]$$

$$+ 2[w_1 S^{(4)}(t_1^+) + v_1^1 S'(t_1)][g'(t_1) - S'(t_1)]$$

$$+ 2[w_{n-1} S'''(t_n^-) + v_n^2 S''(t_n)][g''(t_n) - S''(t_n)]$$

$$- 2[w_{n-1} S^{(4)}(t_n^-) - v_n^1 S'(t_n)][g'(t_n) - S'(t_n)].$$

Looking at the last six lines, it follows from the results of Theorems 4.3.1 and 4.3.2 that the left-hand factors in the two summations are zero for all end condition types. If Type I end conditions are used, then $g'(t_1) = S'(t_1)$, $g'(t_n) = S'(t_n)$, $g''(t_1) = S''(t_1)$, and $g''(t_n) = S''(t_n)$. Thus, the right-hand factors in the last four lines are zero. If Type II end conditions are used, then using the fact that $S(t)$ is linear outside of $[t_1, t_n]$, and the results from Theorems 4.3.1 and 4.3.2, it follows that the left-hand bracketed factors in the last four lines are zero. If Type III end conditions are used, then $g'(t_1) = g'(t_n)$, $g''(t_1) = g''(t_n)$, $S'(t_1) = S'(t_n)$, and $S''(t_1) = S''(t_n)$. Thus combining lines three and five, and lines four and six, we have

$$2\big[w_{n-1} S'''(t_n^-) - w_1 S'''(t_1^+) + v_1^2 S''(t_1) + v_n^2 S''(t_n)\big]\big[g''(t_1) - S''(t_1)\big]$$

and

$$2\big[w_1 S^{(4)}(t_1^+) - w_{n-1} S^{(4)}(t_n^-) + v_1^1 S'(t_1) + v_n^1 S'(t_n)\big]\big[g'(t_1) - S'(t_1)\big].$$

The middle factor in both of these terms is zero because of the Type III end conditions imposed on $S(t)$. Therefore, $V(g) - V(S) - V(g - S) = 0$, and thus we have shown that $V(g) = V(S) + V(g - S)$. Since $V(g - S) \geq 0$, it follows that $V(S) \leq V(g)$ for all $g(t)$ in $H^3[t_1, t_n]$.

## 4.4.  Shape Control Properties of the Weighted Tau-Spline

In this section, we first investigate the effects of the shape control parameters $v_i^1, v_i^2$, and $w_i$ seperately and then in combinations. Throughout this section, we will assume that all three parameter values default to $w_i = 1$, $v_i^1 = 0$, and $v_i^2 = 0$ unless otherwise stated.

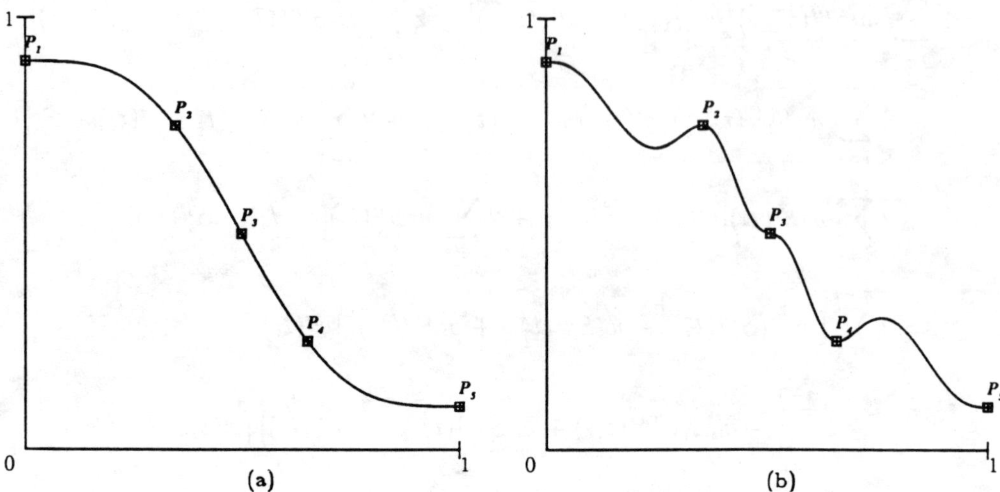

FIG. 4.1. *The functional weighted $\tau$-spline does not mimic splines in tension well. For each curve,$y_1'=y_5'=0$ and $y_1''=y_5''=0$. (a) A natural quintic spline function. (b) $v_2^1=v_3^1=v_4^1=1000000$.*

In analyzing the effects of the tension parameter $v_i^1$ on $S(t)$, we turn to the equations in (4.2). If we keep the adjacent interval weights bounded and divide through by $v_i^1$, then in the limit as $v_i^1 \to \infty$, we find that $S'(t_i)$ approaches 0. In the functional case, this causes the curve to approach either a relative minimum or maximum point, or a point of inflection at $t = t_i$. Figure 4.1 shows that this can lead to an undesireable curve shape. The "nice looking" functional curve in Fig. 4.1a is the natural $C^4$ quintic spline with $y_1' = y_5' = 0$, and $y_1'' = y_5'' = 0$. By setting $v_2^1 = v_3^1 = v_4^1 = 1000000$ in Fig. 4.1b, we see that the distorted curve appears to "flatten out" at the points $\mathbf{P}_2$ (a relative maximum), $\mathbf{P}_3$ (a point of inflection), and $\mathbf{P}_4$ (a relative minimum). In contrast, the "splines under tension" of Schweikert [20] used a global tension parameter which as it increased, the curve converged to the straight line segments connecting the data points. The "flattening" effect was noted by Nielson [15],[16] which led him to comment that $\nu$-splines in the functional case are not very interesting since they do not mimic splines under tension very well. However, in the parametric case, $\mathbf{S}(t)$ has some very interesting properties. Recall that with $\nu$-splines, when two adjacent tension values where increased, the curve converged to the straight line segment connecting the two points $(x_i, y_i)$ and $(x_{i+1}, y_{i+1})$. In the case of weighted $\tau$-splines, when two adjacent tension parameters $v_i^1$ and $v_{i+1}^1$ are increased to apply tension to the curve segment, the graph of $\mathbf{S}(t)$, $t_i \leq t \leq t_{i+1}$, moves *beyond* the straight line segment that one might at first expect. Figure 4.2 shows a family of parametric weighted $\tau$-splines interpolating to the three corners of an equilaterial triangle using periodic end conditions and unit parameterization. As the tension parameters $v_i^1$ are increased, the first derivatives approach the zero vector with the curve approaching what appears to be a *cusp* at the data points.

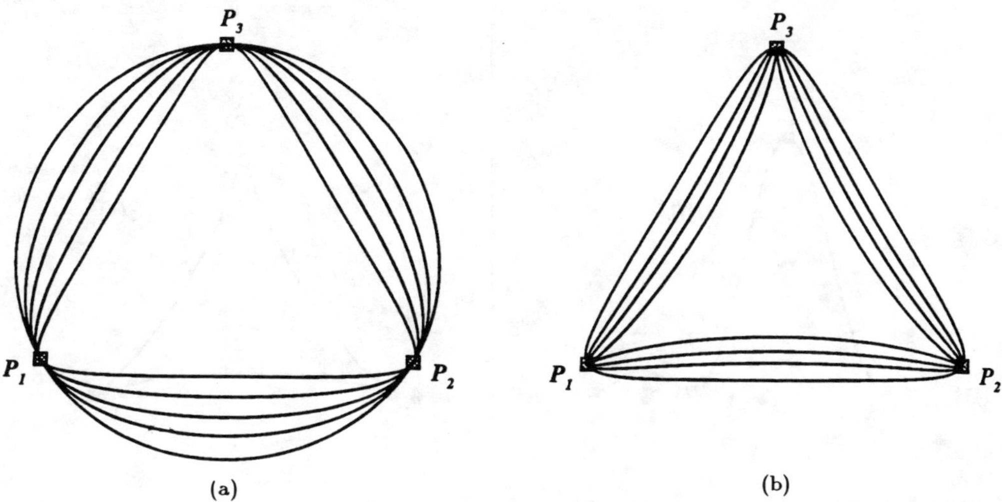

FIG. 4.2. *The weighted $\tau$-spline with increasing values of $v_i^1$. Starting from the outside curves and moving in* (a) $v_i^1 = 0$, 30, 65, 130, *and* 252, *respectively, at each point.* (b) $v_i^1 = 253$, 500, 1000, *and* 1000000, *respectively, at each point.*

For the parametric curve $(x(t), y(t))$, Epstein [7] shows that if $x'(\xi) = y'(\xi) = 0$ but either $x''(\xi) \neq 0$ or $y''(\xi) \neq 0$, then the curve will have a cusp at the point $(x(\xi), y(\xi))$. To see if the weighted $\tau$-spline curve is actually approaching a cusp at the data points, we analyze the two component functions to verify that they do not have simultaneous points of inflection at points where their first derivatives vanish. The cross-plot in Fig. 4.3 shows that with $v_i^1$ increased to 1000000 at each point, the only point at which we have both a horizontal tangent and a point of inflection is at $(2, x(2))$. However, the second derivative at $(2, y(2))$ is not zero.

To analyze the effects of $v_i^2$ on $S(t)$, we turn to the equations in (4.3). If we keep the adjacent interval weights bounded and divide through by $v_i^2$, then as $v_i^2 \to \infty$, we find that $S''(t_i)$ approaches 0 indicating that $S(t)$ becomes more locally linear at $t = t_i$. This is the case for both functional and parametric data. Figure 4.4 shows the effect of increasing values of $v_i^2$ at each corner of an equilateral triangle. Notice the minimal change in the curve when $v_i^2$ is increased from 100 to 1000000.

If $S(t)$ has interval weights $w_i = c$ where $c > 0$, and nonnegative point tensions $v_i^1$ and $v_i^2$, then $S(t)$ reduces to Hagen's $\tau$-spline with point tensions $v_i^1/c$ and $v_i^2/c$. This can be seen by dividing the equations in (4.2) and (4.3) and the end condition equations in (4.4) through by $c$, leaving all interval weights of 1. If we look at $S(t)$ starting with its default parameter values and allow all the interval weights to increase at the same rate, the shape of the curve remains unchanged. In order to change the shape of the curve on the $i$th interval, $w_i$ must change relative to its adjacent neighbors. Thus in Fig. 4.5, we hold $w_1$ and $w_3$ fixed at 1 while $w_2$ takes on the increasing values of 1, 10, and 100. There is very little change in the curve for values of $w_2$ greater than 100. The

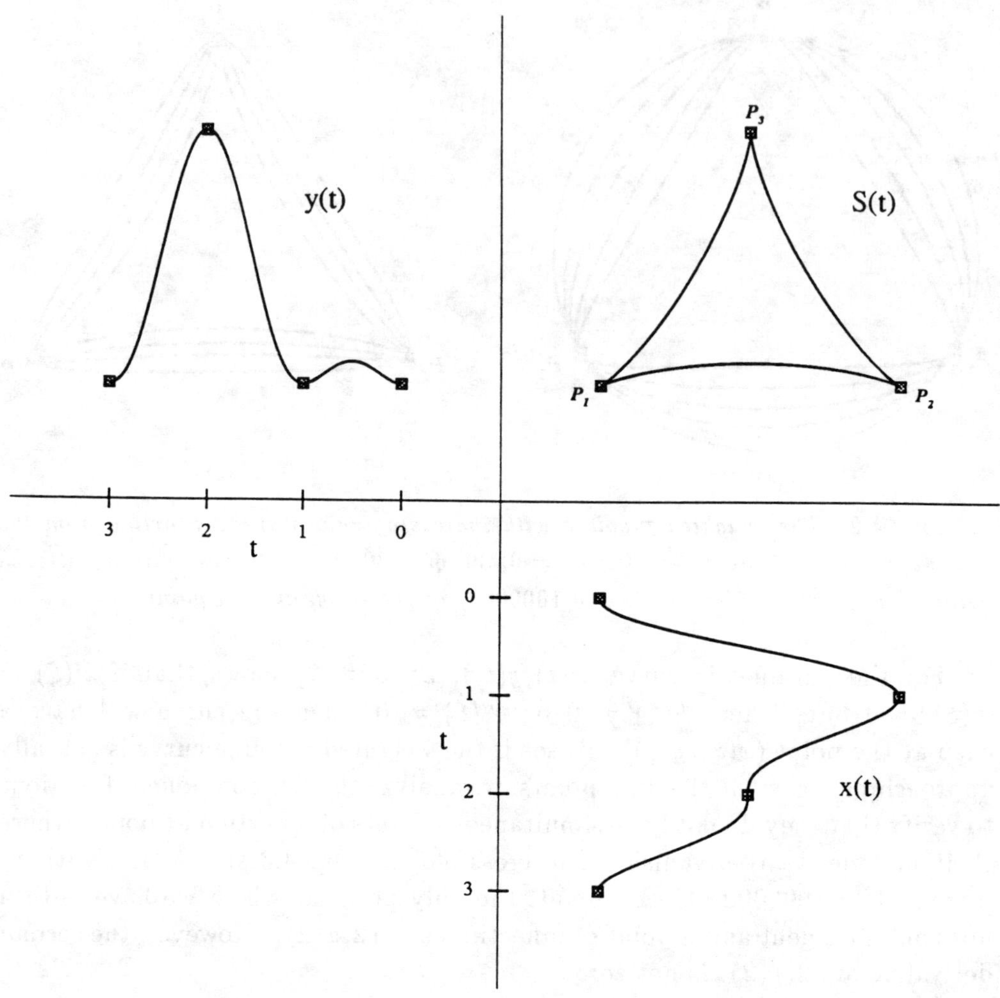


FIG. 4.3. *A cross-plot of the parametric curve* $S(t)$ *in Fig.* 4.2b *with* $v_i^1 = 1000000$ *for all* $i$. *The functional component curves* $(t, y(t))$ *and* $(t, x(t))$ *are shown in the upper left- and lower right-hand corners, respectively.*

same three curves are generated if $w_2$ is fixed at 1 and both $w_1$ and $w_3$ take on the decreasing values of 1, 0.1, and 0.01. In general, if $w_i$ increases without bound while $w_{i-1}, w_{i+1}, v_i^1, v_{i+1}^1, v_i^2$, and $v_{i+1}^2$ remain bounded, then $S'''(t)$ approaches 0 on $(t_i, t_{i+1})$, which means that $S(t)$ is approaching a quadratic segment on $(t_i, t_{i+1})$. The following theorem formalizes this limiting behavior.

THEOREM 4.4.1. *Let* $S(t)$ *be the Type* II *weighted* $\tau$-*spline interpolant to the data* $(t_i, y_i)$, $i = 1, \cdots, n$, *with interval weights* $w_i$, $i = 1, \cdots, n-1$, *and point tension factors* $v_i^1$ *and* $v_i^2$, $i = 1, \cdots, n$. *If* $w_{k-1}, w_{k+1}, v_k^1, v_{k+1}^1, v_k^2$, *and* $v_{k+1}^2$ *are bounded for a given* $k = 1, \cdots, n-1$ *and* $w_k \to \infty$, *then*

$$\lim_{w_k \to \infty} S(t) = (t_{k+1} - t)h_k^{-1}y_k + (t - t_k)h_k^{-1}y_{k+1} + \tfrac{1}{2}(t - t_k)(t - t_{k+1})y_k'',$$

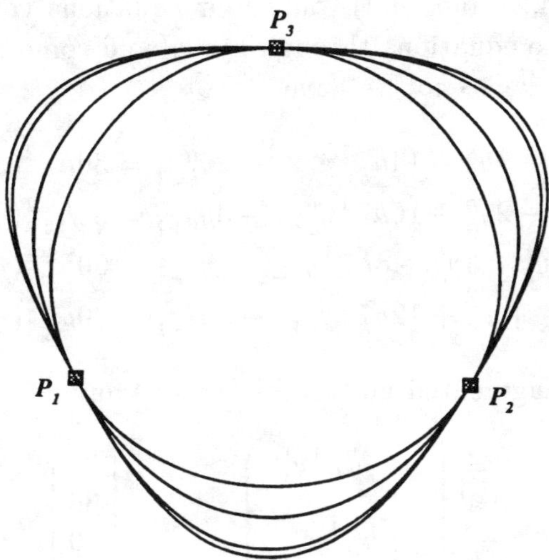

FIG. 4.4. *The weighted $\tau$-spline with increasing values of $v_i^2$. Looking at the curves starting from the inside and moving out, $v_i^2 = 0, 10, 100$, and $1000000$, respectively, at each point. As the $v_i^2 s$ approach $\infty$, the spline curves become more locally linear at the interpolating points.*

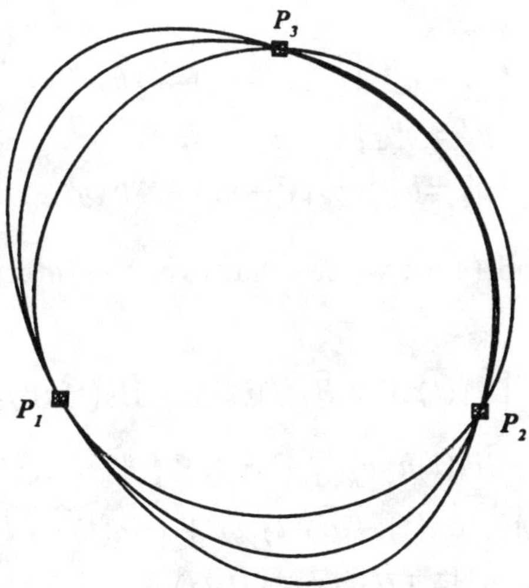

FIG. 4.5. *The weighted $\tau$-spline with increasing values of $w_2$. Looking at the second curve segment and starting with the outside curve segment and moving in, $w_2 = 1, 10$, and $100$. As $w_2$ approaches $\infty$, the spline curve approaches a quadratic segment between the points $P_2$ and $P_3$.*

*for* $t \in (t_k, t_{k+1})$.

*Proof.* From (4.2) and (4.3), only four equations contain the factor $w_k$. After dividing these equations through by $w_k$ and some common factors, and taking the limit as $w_k \to \infty$, we have

$$16h_k^{-1}y_k' + 3y_k'' + 14h_k^{-1}y_{k+1}' - 2y_{k+1}'' = 30h_k^{-2}(y_{k+1} - y_k),$$
$$14h_k^{-1}y_k' + 2y_k'' + 16h_k^{-1}y_{k+1}' - 3y_{k+1}'' = 30h_k^{-2}(y_{k+1} - y_k),$$
$$12h_k^{-1}y_k' + 3y_k'' + 8h_k^{-1}y_{k+1}' - y_{k+1}'' = 20h_k^{-2}(y_{k+1} - y_k),$$
$$8h_k^{-1}y_k' + y_k'' + 12h_k^{-1}y_{k+1}' - 3y_{k+1}'' = 20h_k^{-2}(y_{k+1} - y_k).$$

Reducing the augmented matrix for the system

$$\begin{bmatrix} 16 & 3 & 14 & -2 \\ 14 & 2 & 16 & -3 \\ 12 & 3 & 8 & -1 \\ 8 & 1 & 12 & -3 \end{bmatrix} \cdot \begin{bmatrix} h_k^{-1}y_k' \\ y_k'' \\ h_k^{-1}y_{k+1}' \\ y_{k+1}'' \end{bmatrix} = \begin{bmatrix} 30 \\ 30 \\ 20 \\ 20 \end{bmatrix} h_k^{-2}(y_{k+1} - y_k)$$

we find that

$$\begin{bmatrix} 16 & 3 & 14 & -2 & 30 \\ 14 & 2 & 16 & -3 & 30 \\ 12 & 3 & 8 & -1 & 20 \\ 8 & 1 & 12 & -3 & 20 \end{bmatrix} \longrightarrow \begin{bmatrix} 16 & 3 & 14 & -2 & 30 \\ 0 & -1 & 6 & -2 & 6 \\ 0 & 0 & 2 & -1 & 2 \\ 0 & 0 & 0 & 0 & 0 \end{bmatrix}.$$

After back substituting, it follows that

(4.6)
$$\begin{aligned} y_{k+1}' &= h_k^{-1}(y_{k+1} - y_k) + \tfrac{1}{2}h_k y_{k+1}'', \\ y_k'' &= y_{k+1}'', \\ y_k' &= h_k^{-1}(y_{k+1} - y_k) - \tfrac{1}{2}h_k y_k''. \end{aligned}$$

Thus from (4.1) and (4.6), we see that over the interval $(t_k, t_{k+1})$, that

$$\begin{aligned} \lim_{w_k \to \infty} S(t) &= \lim_{w_k \to \infty} \big[ \widehat{H}_1(t)y_k + \widehat{H}_2(t)y_{k+1} + \widehat{H}_3(t)h_k y_k' \\ &\quad + \widehat{H}_4(t)h_k y_{k+1}' + \widehat{H}_5(t)h_k^2 y_k'' + \widehat{H}_6(t)h_k^2 y_{k+1}'' \big] \\ &= \big[ \widehat{H}_1(t) - \widehat{H}_3(t) - \widehat{H}_4(t) \big] y_k + \big[ \widehat{H}_2(t) + \widehat{H}_3(t) + \widehat{H}_4(t) \big] y_{k+1} \\ &\quad + h_k^2 \big[ -\tfrac{1}{2}\widehat{H}_3(t) + \tfrac{1}{2}\widehat{H}_4(t) + \widehat{H}_5(t) + \widehat{H}_6(t) \big] y_k'' \\ &= (t_{k+1} - t)h_k^{-1}y_k + (t - t_k)h_k^{-1}y_{k+1} + \tfrac{1}{2}(t - t_k)(t - t_{k+1})y_k'' \end{aligned}$$

which is at most a quadratic on $(t_k, t_{k+1})$.

We find that with the weighted $\tau$-spline, there are several combinations of interval weights and point tensions that can tighten the curve down to a straight line segment between consecutive data points. For parametric data, as $v_i^1, v_{i+1}^1, v_i^2$, and $v_{i+1}^2$ get infinitely large, $\mathbf{S}(t)$ approaches the straight line segment between $(x_i, y_i)$ and $(x_{i+1}, y_{i+1})$. This is not the case for functional data. From (4.2) and (4.3), we find only four equations each of which contains exactly one of the factors $v_i^1, v_{i+1}^1, v_i^2$, or $v_{i+1}^2$. Dividing each of these equations through by its respective factor and then taking the limit as that factor goes to infinity, we find that

$$\mathbf{y}_k' = \mathbf{y}_{k+1}' = \mathbf{y}_k'' = \mathbf{y}_{k+1}'' = 0.$$

Thus from (4.1), $\mathbf{S}(t)$ simplifies to

$$\mathbf{S}(t) = \widehat{H}_1(t)\mathbf{y}_i + \widehat{H}_2(t)\mathbf{y}_{i+1}$$
$$= \left[1 - \widehat{H}_2(t)\right]\mathbf{y}_i + \widehat{H}_2(t)\mathbf{y}_{i+1}.$$

Since $\widehat{H}_2(t)$ varies from 0 to 1 as $t$ varies from $t_k$ to $t_{k+1}$, $\mathbf{S}(t)$ traces out the straight line segment from $(x_i, y_i)$ to $(x_{i+1}, y_{i+1})$.

In Fig. 4.6, using the same equilaterial triangle as before, we generate a family of curves where $v_2^1 = v_3^1 = 0, 100, 1000, 10000$, while $v_2^2 = v_3^2 = 0, 2, 20, 200$ resulting in the second curve segment approaching a linear segment. Notice that as we increase $v_2^1, v_2^2, v_3^1$, and $v_3^2$, the bend in the curve at the second and third data points becomes sharper and the curve becomes somewhat tighter on the first and third intervals. Using larger values for the $v_i^1$s and $v_i^2$s resulted in no appreciable change in the shape of the curve. Notice that this "corner" effect is similar to what one gets with the $\nu$-spline for increasing values of $v_i$. To prevent the curve segment in Fig. 4.6 from passing beyond the straight line segment during the convergence, we could have chosen $v_2^1$ and $v_2^2$ such that $v_2^1 = 20v_2^2 + 144$, and $v_3^1$ and $v_3^2$ such that $v_3^1 = 20v_3^2 + 144$. Just where this data-dependent convexity-preserving linear relationship between $v_i^1$s and $v_i^2$s came from is discussed in the Bézier section of [14].

Combinations of $v_i^1, v_{i+1}^1$, and $w_i$ can also be used to force $S(t)$ to approach the line segment joining $(t_i, y_i)$ and $(t_{i+1}, y_{i+1})$. This is formalized in the following theorem.

THEOREM 4.4.2. *Let $S(t)$ be the Type II weighted $\tau$-spline interpolant to the data $(t_k, y_k)$, $k = 1, \cdots, n$, with interval weights $w_k$, $k = 1, \cdots, n-1$, and point tension factors $v_k^1$ and $v_k^2$, $k = 1, \cdots, n$. If $w_{k-1}, w_{k+1}, v_k^2$, and $v_{k+1}^2$ are bounded for a given $k = 1, \cdots, n-1$ and $av_k^1 + b = cv_{k+1}^1 + d = w_k \to \infty$ where $a > 0$ and $c > 0$, then $S(t)$ approaches a straight line segment on the kth interval.*

*Proof.* From (4.2) and (4.3), only four equations contain the factors $v_k^1$, $v_{k+1}^1$, or $w_k$. After dividing these equations through by $w_k$ and some common factors, and taking the limit as $w_k \to \infty$, we have

$$(192 + h_k^3/a)y_k' + 36h_k y_k'' + 168y_{k+1}' - 24h_k y_{k+1}'' = 360h_k^{-1}(y_{k+1} - y_k)$$
$$12y_k' + 3h_k y_k'' + 8y_{k+1}' - h_k y_{k+1}'' = 20h_k^{-1}(y_{k+1} - y_k)$$
$$168y_k' + 24h_k y_k'' + (192 + h_k^3/c)y_{k+1}' - 36h_k y_{k+1}'' = 360h_k^{-1}(y_{k+1} - y_k)$$
$$8y_k' + h_k y_k'' + 12y_{k+1}' - 3h_k y_{k+1}'' = 20h_k^{-1}(y_{k+1} - y_k).$$

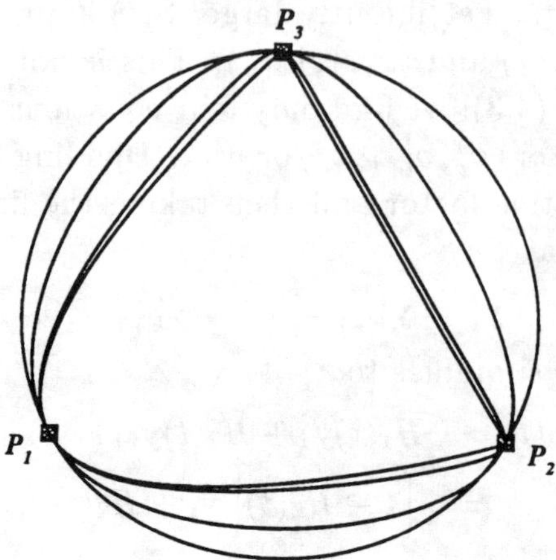


FIG. 4.6. *The weighted $\tau$-spline interpolants using periodic end conditions, unit parameterization, and point tension values of $v_2^1=v_3^1=0$ and $v_2^2=v_3^2=0$, $v_2^1=v_3^1=100$ and $v_2^2=v_3^2=2$, $v_2^1=v_3^1=1000$ and $v_2^2=v_3^2=20$, and $v_2^1=v_3^1=10000$ and $v_2^2=v_3^2=200$. The resulting curves are in the order from outside in on the second curve segment.*

Solving this system we find

$$
\begin{aligned}
y'_k &= \frac{60a}{h_k(h_k^3 + 30a + 30c)}(y_{k+1} - y_k), \\[2mm]
y'_{k+1} &= \frac{5(h_k^3 + 12c - 12a)}{h_k^2(h_k^3 + 30a + 30c)}(y_{k+1} - y_k), \\[2mm]
y''_k &= \frac{60c}{h_k(h_k^3 + 30a + 30c)}(y_{k+1} - y_k), \\[2mm]
y''_{k+1} &= \frac{-5(h_k^3 - 12c + 12a)}{h_k^2(h_k^3 + 30a + 30c)}(y_{k+1} - y_k).
\end{aligned}
$$

(4.7)

Thus from (4.1) and (4.7), it follows that over the interval $[t_k, t_{k+1}]$,

$$
\begin{aligned}
\lim_{w_k \to \infty} S(t) ={} & \widehat{H}_1(t)y_k + \widehat{H}_2(t)y_{k+1} + \widehat{H}_3(t)\frac{60a(y_{k+1} - y_k)}{h_k^3 + 30a + 30c} \\[2mm]
& + \widehat{H}_4(t)\frac{60c(y_{k+1} - y_k)}{h_k^3 + 30a + 30c} + \widehat{H}_5(t)\frac{5(h_k^3 + 12c - 12a)}{h_k^3 + 30a + 30c}(y_{k+1} - y_k) \\[2mm]
& + \widehat{H}_6(t)\frac{-5(h_k^3 - 12c + 12a)}{h_k^3 + 30a + 30c}(y_{k+1} - y_k) \\[2mm]
={} & [1 - g(t)]\, y_k + g(t)y_{k+1}
\end{aligned}
$$

where

$$
g(t) = \frac{(t - t_k)\,[2(t - t_k)^4 - 5(t - t_k)^3 h_k + 5(t - t_k)h_k^3 - 60(t - t_k)(a - c) + 120ah_k]}{2h_k^2\,[30(a + c) + h_k^3]}.
$$

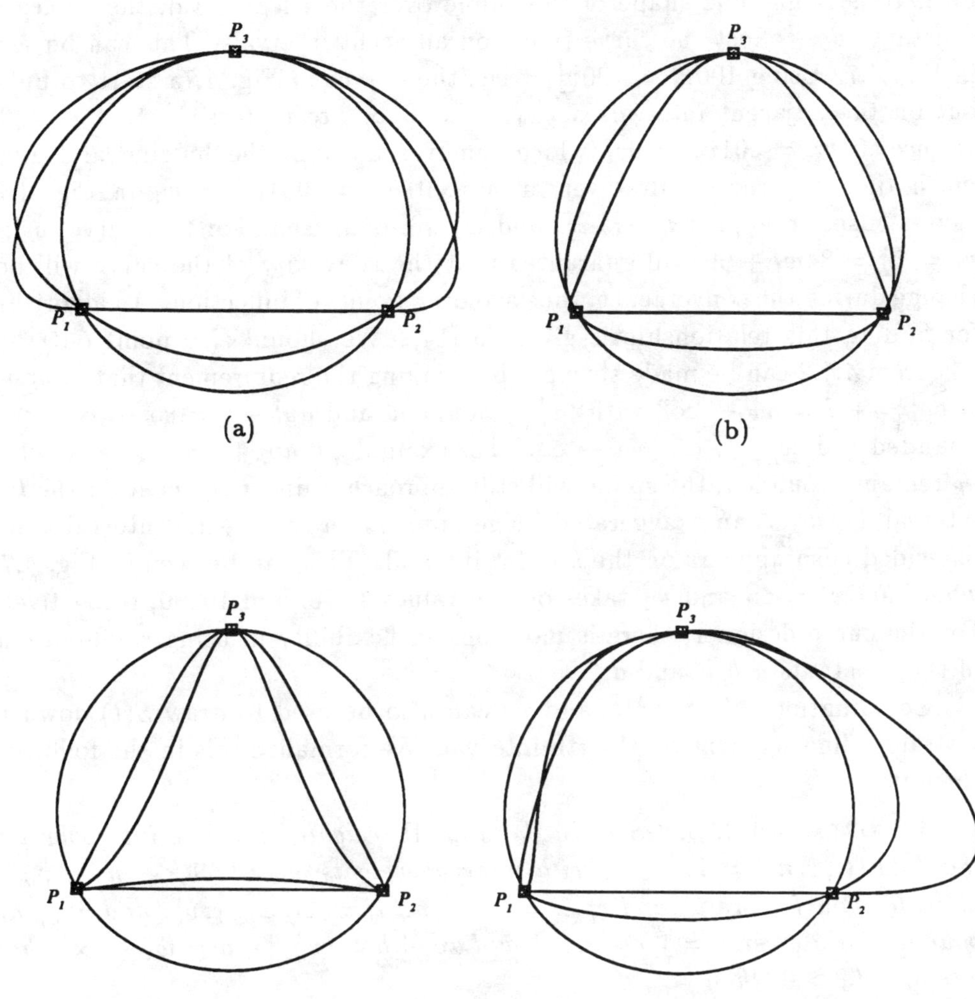



FIG. 4.7. *The weighted $\tau$-spline interpolants using periodic end conditions, unit parameterization, and various combinations of $v_i^1 s$ and $w_i s$.*

Since

$$0 = g(t_k) \leq g(t) \leq g(t_{k+1}) = 1$$

for $t \in [t_k, t_{k+1}]$, it follows that $S(t)$ traces out the straight line segment from $(t_k, y_k)$ to $(t_{k+1}, y_{k+1})$ as $t$ varies from $t_k$ to $t_{k+1}$.

For the function $g(t)$ in Theorem 4.4.2, it is interesting to note that

$$\lim_{a=c\to\infty} g(t) = \frac{t - t_k}{h_k},$$

which is the straight line from $(t_k, 0)$ to $(t_{k+1}, 1)$.

Even though the choice of the constants $a, b, c$, and $d$ does not affect the ultimate straight line shape of the spline over the $i$th interval, the choice of constants *does* affect the curve shape on adjacent intervals. This can be seen in Fig. 4.7. Using $100v_1^1 = 100v_2^1 = w_1$, the curves in Fig. 4.7a start to bulge out on the adjacent intervals as $w_1$, $v_1^1$ and $v_2^1$ are increased. In Fig. 4.7b, we use $.01v_1^1 = .01v_2^1 = w_1$ which removes most of the bulging and gives much tighter corners. However, using $.001v_1^1 = .001v_2^1 = w_1$ in Fig. 4.7c causes cusps to appear as $v_1^1$, $v_2^1$, and $w_1$ are increased. For this curve, using $v_1^1 = v_2^1 = 84w_1 + 60$ will guarantee that the convexity of the curve will not change during the convergence, thus avoiding points of inflection. An algorithm for finding this relationship is given in [14]. We should also point out that Theorem 4.4.2 can be made stronger by relaxing the requirement that "$av_k^1 + b = cv_{k+1}^1 + d = w_k \to \infty$" with "$v_{k+1}^1$ bounded and $av_k^1 + b = w_k \to \infty$, or $v_k^1$ bounded and $cv_{k+1}^1 + d = w_k \to \infty$." For example, if $av_k^1 + b = w_k \to \infty$ while $v_k$ remains bounded, the spline will still approach a linear segment on the $k$th interval, however, an exaggerated bulge appears on the $k + 1$st interval and a one-sided cusp appears on the $k - 1$st interval. This can be seen in Fig. 4.7d where $.001v_1^1 = w_1$ and $w_1$ takes on the values 1, 10, and 10000, respectively. For the curve designer, there is more design flexibility in using combinations of the constants $a, b, c$, and $d$.

Combinations of $v_i^2$, $v_{i+1}^2$, and $w_i$ can also be used to draw $S(t)$ down to a straight line segment on the $i$th interval. We formalize this in the following theorem.

THEOREM 4.4.3. *Let $S(t)$ be the Type* II *weighted $\tau$-spline interpolant to the data* $(t_i, y_i)$, $i = 1, \cdots, n$, *with interval weights* $w_i$, $i = 1, \cdots, n - 1$, *and point tension factors* $v_i^1$ *and* $v_i^2$, $i = 1, \cdots, n$. *If* $w_{k-1}, w_{k+1}, v_k^1$, *and* $v_{k+1}^1$ *are bounded for a given* $k = 1, \cdots, n - 1$ *and* $av_k^2 + b = cv_{k+1}^2 + d = w_k \to \infty$ *where* $a > 0$ *and* $c > 0$, *then*

$$\lim_{w_k \to \infty} S(t) = (t_{k+1} - t)h_k^{-1}y_k + (t - t_k)h_k^{-1}y_{k+1}.$$

*Proof.* From (4.2) and (4.3), only four equations contain the factors $v_k^2$, $v_{k+1}^2$, or $w_k$. After dividing these equations through by $w_k$ and some common factors, and taking the limit as $w_k \to \infty$, we have

$$16y_k' + 3h_ky_k'' + 14y_{k+1}' - 2h_ky_{k+1}'' = 30h_k^{-1}(y_{k+1} - y_k),$$
$$36y_k' + h_k(9 + h_k/a)y_k'' + 24y_{k+1}' - 3h_ky_{k+1}'' = 60h_k^{-1}(y_{k+1} - y_k),$$
$$14y_k' + 2h_ky_k'' + 16y_{k+1}' - 3h_ky_{k+1}'' = 30h_k^{-1}(y_{k+1} - y_k),$$
$$24y_k' + 3h_ky_k'' + 36y_{k+1}' - h_k(9 + h_k/c)y_{k+1}'' = 60h_k^{-1}(y_{k+1} - y_k).$$

Solving this system we find

$$(4.8) \qquad \begin{aligned} y_k' &= y_{k+1}' = h_k^{-1}(y_{k+1} - y_k), \\ y_k'' &= y_{k+1}'' = 0. \end{aligned}$$

Thus from (4.1) and (4.8), it follows that over the interval $[t_k, t_{k+1}]$,

$$\lim_{w_k \to \infty} S(t) = \widehat{H}_1(t)y_k + \widehat{H}_2(t)y_{k+1} + \widehat{H}_3(t)(y_{k+1} - y_k) + \widehat{H}_4(t)(y_{k+1} - y_k)$$

$$= \left[\widehat{H}_1(t) - \widehat{H}_3(t) - \widehat{H}_4(t)\right]y_k + \left[\widehat{H}_2(t) + \widehat{H}_3(t) + \widehat{H}_4(t)\right]y_{k+1}$$

$$= (t_{k+1} - t)h_k^{-1}y_k + (t - t_k)h_k^{-1}y_{k+1}.$$

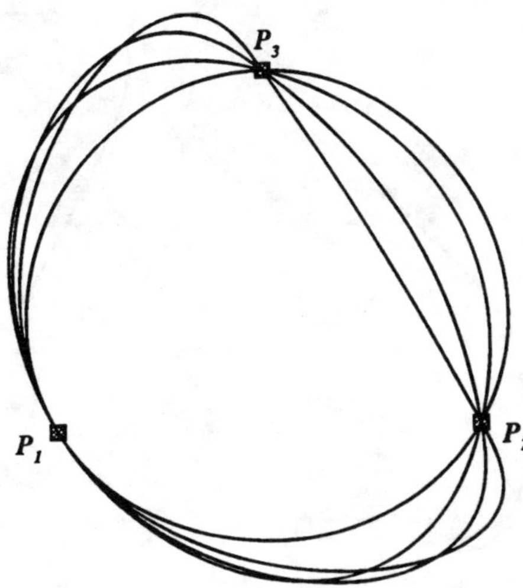


FIG. 4.8. *The weighted $\tau$-spline interpolants using periodic end conditions, unit parameterization, and combinations of $v_2^2$, $v_3^2$, and $w_2$.*

Figure 4.8 shows a sequence of converging curves using $v_2^2 = v_3^2 = w_2 = 0, 10, 100,$ and $1000$. Notice that as $v_2^2$, $v_3^2$, and $w_2$ are increased, the second curve segment goes linear while the adjacent intervals tend to bulge out more and more. There is a minimal change in the shape of the curve for values larger than 1000. We note that Theorem 4.4.3 can be made stronger by requiring only that either $v_{k+1}^2$ be bounded and $a > 0$, or that $v_k^2$ be bounded and $c > 0$. That is, the proof of Theorem 4.4.3 is still valid if either $a = 0$ or $c = 0$, but not both. The choices of the constants $a, b, c,$ and $d$ have little effect on the ultimate shape of the curve on the $i$th interval **or** on the adjacent intervals. This would seem to follow from (4.8). That is, the curves in Fig. 4.8 are visually duplicated using $v_3^2 = 0$, or $v_2^2 = 0$. Different choices of $v_i^2$ have little or no effect on modifying the bulges on adjacent intervals. The straight line shape of $S(t)$ on the $i$th interval depends only on either $v_i^2$ or $v_{i+1}^2$ going to infinity with $w_i$, not both.

The final parameter that affects the shape of the curve is that of parameterization. Since it is still an open research question as to which parameterization yields the "best" shaped curve, we comment only on the effects of a linear reparameterization. For a parametric weighted $\tau$-spline $\mathbf{S}(t)$ whose knot vector is first scaled then translated: if $\mathbf{S} \in C^4$, then the shape of the curve is unchanged under *any* linear parameter transformation; if $\mathbf{S} \in C^3 - C^4$, then the shape of the curve is unchanged if the original $v_i^1$ parameters are divided by the cube of the scaling factor in the transformation; if $\mathbf{S} \in C^2 - C^3$, then the same curve can be generated by inversely scaling the $v_i^1$ and $v_i^2$ parameters by the cube of the scaling factor and by the scaling factor, respectively. A proof for this is given in [14].

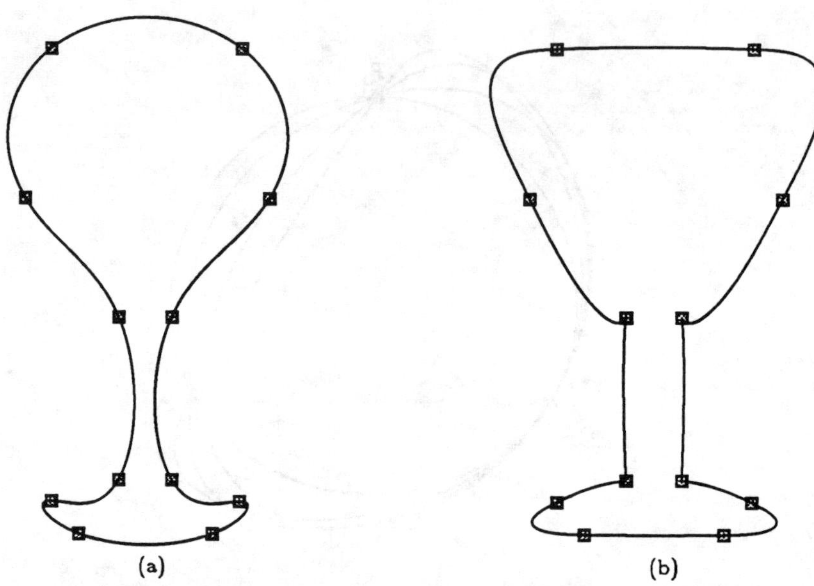

(a)                 (b)

FIG. 4.9. *The parametric weighted $\tau$-spline with unit parameterization and periodic end conditions where* (a) *all parameters are set to default values (the $C^4$ natural quintic spline), and* (b) $v_3^1=v_4^1=v_9^1=v_{10}^1=10^6$, *and* $v_i^1=0$ *otherwise;* $v_4^2=v_5^2=v_8^2=v_9^2=10^6$, $v_6^2=v_{12}^2=1000$, *and* $v_i^2=0$ *otherwise;* $w_6=w_{12}=1000$, *and* $w_i=1$ *otherwise. The points are numbered counterclockwise starting from the upper left-hand corner.*

The parametric curve in Fig. 4.9 takes advantage of several of the curve-shaping properties of the weighted $\tau$-spline. The natural quintic spline is passed through the data in Fig. 4.9a where the points are numbered consecutively starting with the point in the upper left-hand corner and moving counter-clockwise. The wineglass shape in Fig. 4.9b is formed with "visual cusps" at the third and tenth points ($v_3^1 = v_{10}^1 = 10^6$), sharp corners at the fourth and ninth points ($v_4^1 = v_4^2 = v_9^1 = v_9^2 = 10^6$), local flatness at the fifth and eighth points ($v_5^2 = v_8^2 = 10^6$), and interval flatness on the sixth and twelvth intervals with bulging on adjacent intervals ($v_6^2 = w_6 = v_{12}^2 = w_{12} = 1000$). The third and ninth intervals can be made tighter by increasing $v_3^2$ and $v_9^2$, but then the visual cusp effect is diminished.

## 4.5.  Surfaces Using Weighted Tau-Spline Curves

In this section, we generalize to the three-dimensional weighted $\tau$-spline curve and use it with 3D tensor product surface patches defined over the rectangle $[0,1] \times [0,1]$. These patches will then be joined with certain boundary connectivity conditions to form a surface of $m$-by-$n$ biquintic Hermite patches. The weighted $\tau$-spline is also used to generate a $C^2$ biquintic spline-blended surface.

Whereas a parametric curve uses one parameter for its definition, a parametric surface uses two: $t$ and $u$. We generalize the 16-parameter bicubic interpolant of Ferguson [8], which has been popularized by Coons [5], to the 36-parameter biquintic interpolant (see Barnhill [2]). This interpolant requires knowledge of position, two first derivative vectors, two second derivative vectors, and four "twists" vectors at each of the four corner points of the patch. Whereas the position data and the first derivative vectors (and to some degree the second derivative vectors) are useful design tools, the twist vectors generally are not. This is due to the fact that they are more difficult to visualize geometrically. Methods of determining twists have received a great deal of attention [2],[3],[10],[12], but the question as to what twists yield the best looking surface remains unanswered. One easy solution is to set all twist vectors to the zero vector. However, zero twists cause "pseudoflats" which Forrest [10] has characterized as "thumbprints" in the surface.

The 36-parameter biquintic surface patch $\mathbf{S}(t,u)$ defined over the parameter domain $0 \le t, u \le 1$ is given by

$$\mathbf{S}(t,u) = [H_1(t)\ H_2(t)\ H_3(t)\ H_4(t)\ H_5(t)\ H_6(t)]$$

$$(4.9) \quad \cdot
\begin{bmatrix}
\mathbf{S}_{00} & \mathbf{S}_{01} & \mathbf{S}_{00}^{u} & \mathbf{S}_{01}^{u} & \mathbf{S}_{00}^{uu} & \mathbf{S}_{01}^{uu} \\
\mathbf{S}_{10} & \mathbf{S}_{11} & \mathbf{S}_{10}^{u} & \mathbf{S}_{11}^{u} & \mathbf{S}_{10}^{uu} & \mathbf{S}_{10}^{uu} \\
\mathbf{S}_{00}^{t} & \mathbf{S}_{01}^{t} & \mathbf{S}_{00}^{tu} & \mathbf{S}_{01}^{tu} & \mathbf{S}_{00}^{tuu} & \mathbf{S}_{01}^{tuu} \\
\mathbf{S}_{10}^{t} & \mathbf{S}_{11}^{t} & \mathbf{S}_{10}^{tu} & \mathbf{S}_{11}^{tu} & \mathbf{S}_{10}^{tuu} & \mathbf{S}_{10}^{tuu} \\
\mathbf{S}_{00}^{tt} & \mathbf{S}_{01}^{tt} & \mathbf{S}_{00}^{ttu} & \mathbf{S}_{01}^{ttu} & \mathbf{S}_{00}^{ttuu} & \mathbf{S}_{01}^{ttuu} \\
\mathbf{S}_{10}^{tt} & \mathbf{S}_{11}^{tt} & \mathbf{S}_{10}^{ttu} & \mathbf{S}_{11}^{ttu} & \mathbf{S}_{10}^{ttuu} & \mathbf{S}_{10}^{ttuu}
\end{bmatrix}
\cdot
\begin{bmatrix}
H_1(u) \\
H_2(u) \\
H_3(u) \\
H_4(u) \\
H_5(u) \\
H_6(u)
\end{bmatrix}$$

where the $H_k(t)$ and $H_k(u)$, $k = 1, \cdots, 6$, are the univariate quintic Hermite blending functions given in §4.2, and

$$\mathbf{S}(i,j) = \mathbf{S}_{ij}, \qquad \frac{\partial}{\partial u}\left(\frac{\partial \mathbf{S}}{\partial t}\right)(i,j) = \mathbf{S}_{ij}^{tu},$$

$$\frac{\partial \mathbf{S}}{\partial t}(i,j) = \mathbf{S}_{ij}^{t}, \qquad \frac{\partial^2}{\partial u^2}\left(\frac{\partial \mathbf{S}}{\partial t}\right)(i,j) = \mathbf{S}_{ij}^{tuu},$$

$$\frac{\partial \mathbf{S}}{\partial u}(i,j) = \mathbf{S}_{ij}^{u}, \qquad \frac{\partial}{\partial u}\left(\frac{\partial^2 \mathbf{S}}{\partial t^2}\right)(i,j) = \mathbf{S}_{ij}^{ttu},$$

$$\frac{\partial^2 \mathbf{S}}{\partial t^2}(i,j) = \mathbf{S}_{ij}^{tt},$$

$$\frac{\partial^2 \mathbf{S}}{\partial u^2}(i,j) = \mathbf{S}_{ij}^{uu}, \qquad \frac{\partial^2}{\partial u^2}\left(\frac{\partial^2 \mathbf{S}}{\partial t^2}\right)(i,j) = \mathbf{S}_{ij}^{ttuu},$$

where $i, j \in \{0, 1\}$. We should note here that we do not have the compatibility problem as discussed in Barnhill [3] and Gregory [12]. That is, the order in which we compute the mixed partial derivatives is not important. For example,

$$\mathbf{S}_{ij}^{ttu} = \mathbf{S}_{ij}^{tut} = \mathbf{S}_{ij}^{utt}.$$

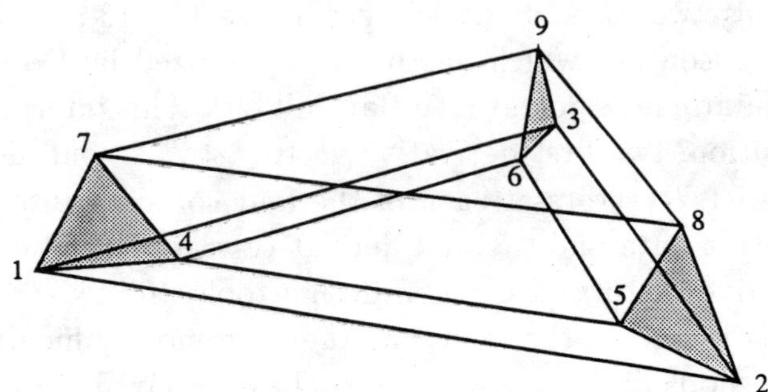

FIG. 4.10. *A triangular three-dimensional control polygon.*

The method that we will use to approximate these twists is the spline-blending technique of Gordon [11]. We begin with a rectangular array of three-dimensional points $\mathbf{P}_{i,j}$ where

$$\mathbf{P}_{i,j} = \begin{pmatrix} x_{i,j} \\ y_{i,j} \\ z_{i,j} \end{pmatrix}$$

for $i = 1, \cdots, m$ and $j = 1, \cdots, n$. The 2D array structure is used to determine order of connectivity in the data points. We next form a network of $C^2$ weighted $\tau$-spline curves which interpolate the data. Given the knot sequence $t_1 < \cdots < t_m$, then for a fixed $j$, $j = 1, \cdots, n$, let $f_j(t)$ be the weighted $\tau$-spline with any one of the three types of end conditions, and point tension values $vt_{i,j}^1 \geq 0$ and $vt_{i,j}^2 \geq 0$ for $i = 1, \cdots, m$, and interval weights $wt_{i,j}^1 > 0$ and $wt_{i,j}^2 > 0$, $i = 1, \cdots, m - 1$, which satisfies

$$f_j(t_i) = \mathbf{P}_{i,j}.$$

Similarly, given the knot sequence $u_1 < \cdots < u_n$, then for each $i$, $i = 1, \cdots, m$, let $g_i(u)$ be the weighted $\tau$-spline with any one of the three types of end conditions, and point tension values $vu_{i,j}^1 \geq 0$ and $vu_{i,j}^2 \geq 0$ for $j = 1, \cdots, n$, and interval weights $wu_{i,j}^1 > 0$ and $wu_{i,j}^2 > 0$, $j = 1, \cdots, n - 1$, which satisfies

$$g_i(u_j) = \mathbf{P}_{i,j}.$$

We now define the two sets of univariate functions which will be used to blend the network of weighted $\tau$-spline curves. Let $c_i(t)$ be the $C^4$ natural quintic spline that satisfies

$$c_i(t_q) = \begin{cases} 1, & \text{if } i = q; \\ 0, & \text{if } i \neq q \end{cases}$$

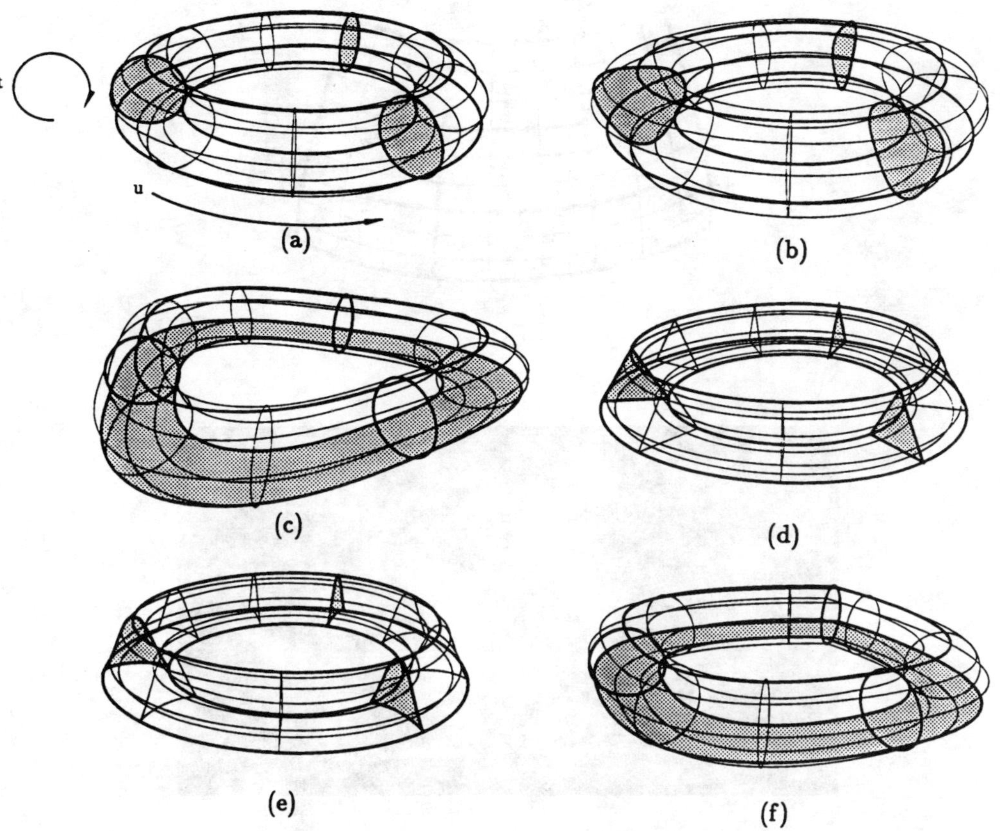



FIG. 4.11. *Spline-blended network of curves through the data shown in Fig. 4.10.* (a) *All parameters are set to default values in both the $t$ and $u$ directions except that,* (b) $v_i^2 = 10^6$ $\forall i$ *in the $t$ direction,* (c) $v_i^2 = 10^6$ $\forall i$ *in the $u$ direction,* (d) $v_i^1 = v_i^2 = 10^6$ $\forall i$ *in the $t$ direction,* (e) $v_1^1 = v_2^1 = 10^6$, *and $v_3^2 = 10^6$ in the $t$ direction,* (f) $w_1^1 = 10^6$, *and $v_3^1 = v_3^2 = 10^6$ in the $u$ direction.*

for $i, q = 1, \cdots, m$. Similarly, let $d_j(u)$ be the $C^4$ natural quintic spline that satisfies

$$d_j(u_r) = \begin{cases} 1, & \text{if } j = r; \\ 0, & \text{if } j \neq r \end{cases}$$

for $j, r = 1, \cdots, n$.

Using the functions defined above, a spline-blended surface is given by

$$(4.10) \qquad B(t,u) = \sum_{i=1}^{m} c_i(t)g_i(u) + \sum_{j=1}^{n} d_j(u)f_j(t) - \sum_{i=1}^{m}\sum_{j=1}^{n} \mathbf{P}_{i,j}c_i(t)d_j(u)$$

where $i = 1, \cdots, m$, and $j = 1, \cdots, n$. The spline-blended surface $B(t,u)$ in (4.10) is a $C^2$ biquintic surface which interpolates to the given network of curves. That is,

$$B(t, u_j) = f_j(t) \qquad \text{for } j = 1, \cdots, n \qquad \text{and}$$
$$B(t_i, u) = g_i(u) \qquad \text{for } i = 1, \cdots, m.$$

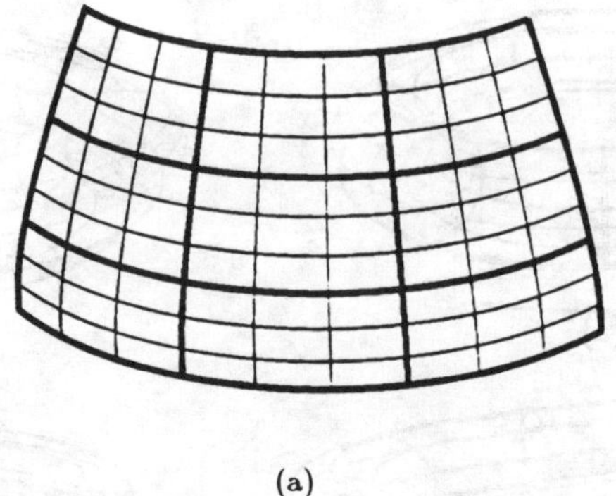

(a)

(b)

FIG. 4.12. *A biquintic Hermite surface through the teapot data. All parameters are set to default values in both the t and u directions.* (a) *Nine surface patches forming a portion of the teapot body with isoparametric lines at 1/3 and 2/3 (network lines in dark), and* (b) *the rendered biquintic Hermite surface.*

We can now compute the four twist vectors at each of the four corner points for the Hermite patches given in (4.9). For example, at the surface knot $(t_q, u_r)$, where $q = 1, \cdots, m$ and $r = 1, \cdots, n$, we have the twists

$$B^{tuu}(t_q, u_r) = \sum_{i=1}^{m} c_i'(t_q)g_i''(u_r) + \sum_{j=1}^{n} d_j''(u_r)f_j'(t_q) - \sum_{i=1}^{m}\sum_{j=1}^{n} \mathbf{P}_{i,j}c_i'(t_q)d_j''(u_r).$$

$B^{tu}(t_q, u_r)$, $B^{ttu}(t_q, u_r)$, and $B^{ttuu}(t_q, u_r)$ are computed in a similar way.

In Fig. 4.10, a nine-point triangular three-dimensional control polygon is given through which we will pass a doubly periodic spline-blended surface. The twist vectors will be computed using the spline-blending function $B(t, u)$ with the blending functions $c_i(t)$ and $d_j(t)$ having periodic end conditions.

We now fit spline-blended surfaces with a variety of tension parameters to the data in Fig. 4.11. In each of the figures in Fig. 4.11, a wireframe of the surface is shown with heavy lines indicating the boundary curves and thin

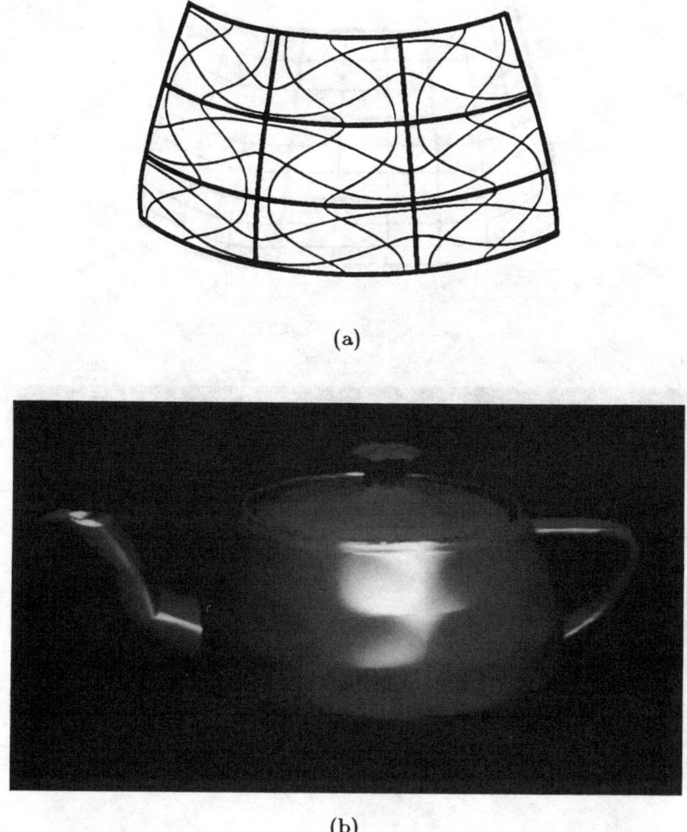

(a)

(b)

FIG. 4.13. *A biquintic Hermite surface through the teapot data. All parameters are set to default values except for* $vu^1_{i,j} = vu^2_{i,j} = 10^6$ *for all* $i+j$ *even, and* $vt^1_{i,j} = vt^2_{i,j} = 10^6$ *for all* $i+j$ *odd. (a) Nine surface patches forming a portion of the teapot body with isoparametric lines at 1/3 and 2/3 (network lines in dark), and (b) the rendered biquintic Hermite surface.*

lines indicating isoparametric lines at parameter values of 1/3 and 2/3 in both directions. The default surface, where $vt^1_i = vt^2_i = vu^1_i = vu^2_i = 0$ and $wt_i = wu_i = 1$ for all meaningful $i$, is shown in Fig. 4.11a with the $t$ and $u$ directions indicated. The point tensions and interval weights for the networks of $\tau$-spline curves are given in the figure caption.

A more complicated object is Martin Newell's well-known Utah teapot [6]. In the $t$ direction, we use natural end conditions, and in the $u$ direction, we use closed periodic end conditions. Above each of the photos in Figs. 4.12–4.14, we show nine surface patches forming a portion of the front of the teapot body. Through each of the nine patches, the isoparametric lines at 1/3 and 2/3 are drawn. We might think of these isoparametric lines as being "stretch marks" on the surface. The network curves are shown in dark. The surfaces shown in the photos are biquintic Hermite surfaces with zero twists. In Fig. 4.14, the

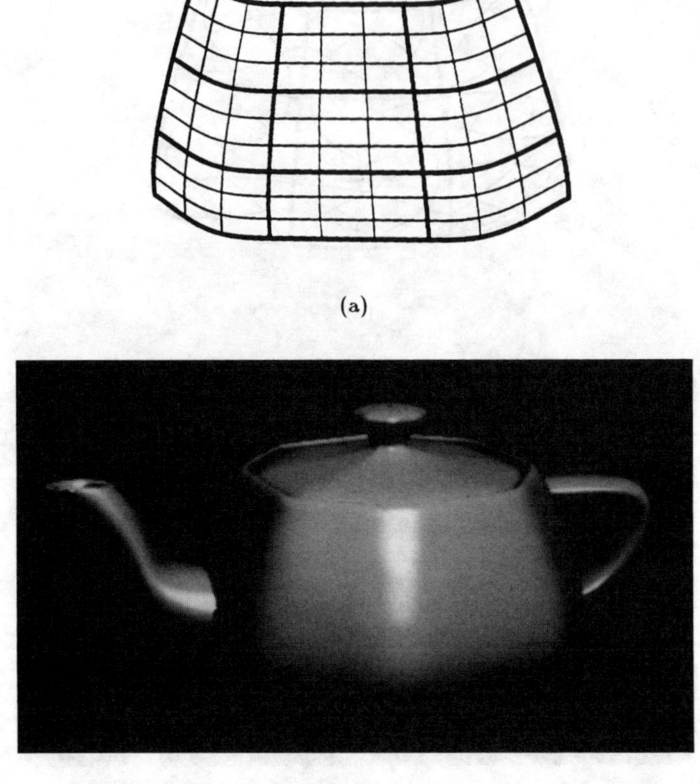

(a)

(b)

FIG. 4.14. *A biquintic Hermite surface through the teapot data. All parameters are set to default values except for* $wu_{3i,j} = vu^2_{3i,j} = 10^6$ *for all meaningful* $i,j$. (a) *Nine surface patches forming a portion of the teapot body with isoparametric lines at* 1/3 *and* 2/3 *(network lines in dark), and* (b) *the rendered biquintic Hermite surface.*

spout and handle were rendered with default parameter values.

## 4.6. Acknowledgments

Some conclusions for curves, along with Fig. 4.9 and parts of Fig. 4.7, were first published in "Quintic spline interpolation witth shape control parameters," pp. 259–262 in *Approximation Theory* VI: *Vol.* 1, Academic Press, New York, 1989.

## References

[1]  J. H. Ahlberg, E. N. Nilson, and J. L. Walsh, *The Theory of Splines and Their Applications*, Academic Press, New York, 1967.

[2]  R. E. Barnhill, *Computer aided surface representation and design*, in Surfaces in CAGD, R. E. Barnhill and W. Boehm, eds., North-Holland, New York, 1983, pp. 1–24.

[3]  R. E. Barnhill, J. H. Brown, and I. M. Klucewicz, *A new twist in computer*

*aided geometric design*, Comput. Graph. Image Process., 1 (1978), pp. 34–47.

[4]  B. A. Barsky, *Exponential and polynomial methods for applying tension to an interpolating spline curve*, Comput. Vision Graph. Image Process., 27 (1984), pp. 1–18.

[5]  S. A. Coons, *Surfaces for Computer Aided Design of Space Forms*, MIT Project MAC TR-41, June 1967.

[6]  F. C. Crow, *The origins of the teapot*, Computer Graphics, 7 (1987), pp. 8–19.

[7]  M. P. Epstein, *On the influence of parameterization in parametric interpolation*, SIAM J. Numer. Anal., 13 (1976), pp. 261–268.

[8]  J. C. Ferguson, *Multivariable curve interpolation*, J. Assos. Comput. Mach., II (1964), pp. 221–228.

[9]  T. A. Foley, *Interpolation with interval and point tension controls using cubic weighted $\nu$-splines*, ACM Trans. Math. Software, 3 (1987), pp. 68–96.

[10]  A. R. Forrest, *The Computation of Bicubic Twist Terms*, Computer-Aided Design Group Document 20, Cambridge University, 1969.

[11]  W. J. Gordon, *Spline-blended surface interpolation through curve networks*, J. Math. Mech., 18 (1969), pp. 931–952.

[12]  J. A. Gregory, *Smooth interpolation without twist constraints*, in Computer Aided Geometric Design, R. E. Barnhill and R. F. Riesenfeld, eds., Academic Press, New York, 1974, pp. 71–87.

[13]  H. Hagen, *Geometric spline curves*, Comput. Aided Geom. Des., 2 (1985), pp. 223–227.

[14]  D. A. Neuser, *Curve and Surface Interpolation using Quintic Weighted Tau-Splines.* Masters Thesis, Comp. Sci. Dept., Ariz. State Univ., 1988.

[15]  G. M. Nielson, *Some piecewise polynomial alternatives to splines under tension*, in Computer Aided Geometric Design, R. E. Barnhill and R. F. Riesenfeld, eds., Academic Press, New York, 1974, pp. 209–235.

[16]  ——, *Rectangular $\nu$-splines*, IEEE Trans. Comput. Graph., 6 (1986), pp. 35–40.

[17]  D. T. Pilcher, *Smooth Approximation of Parametric Curves and Surfaces.* Ph.D. Thesis, Mathematics Department, Univ. of Utah, 1973.

[18]  ——, *Smooth parametric surfaces*, in Computer Aided Geometric Design, R. E. Barnhill and R. F. Riesenfeld, eds., Academic Press, New York, 1974, pp. 237–253.

[19]  K. Salkauskas, $C^1$ *splines for interpolation of rapidly varying data*, Rocky Mountain J. Math., 14 (1984), pp. 239–250.

[20]  D. G. Schweikert, *An interpolation curve using a spline in tension*, J. Math. Phys., 45 (1966), pp. 312–317.

# Weighted Splines Based on Piecewise Polynomial Weight Functions

L. Bos and K. Salkauskas

## 5.1. Introduction

By a weighted spline we mean a spline interpolant which minimizes a weighted semi-norm. Such splines were introduced by Salkauskas [8], where the weight function is piecewise constant on the same partition as the interpolation points. They have proven to be useful in the interpolation of rapidly varying data by $C^1$ piecewise cubics and have, for instance, been successfully exploited by Foley [4]–[6], both in combination with Nielson's $\nu$-splines and in a bivariate analogue of tensor product interpolation.

In [9], we allow for any weight function which is piecewise constant on an arbitrary partition (not necessarily that determined by the interpolation points). As it turns out, such optimal interpolants are also piecewise cubic. Further, in [1] we have shown how to compute optimal interpolants based on general weight functions by approximating the weight by a sequence of piecewise constant weight functions. In this work we consider the analogous results for weights whose *reciprocals* are piecewise polynomial.

## 5.2. Existence of Optimal Weighted Splines

In this section we show that for quite general weight functions there exist optimal interpolating splines. Furthermore, if the reciprocal of the weight function is piecewise polynomial, the optimal spline is also piecewise polynomial. Our approach is reminiscent of that used by Meinguet [7] in the construction of thin-plate splines. An appropriate setting for this analysis is the Sobolev space of the following definition.

DEFINITION 5.2.1. For $k \geq 1$, let $H_k$ be the Sobolev space

$$H_k := \{v \in C^{k-1}(\mathbf{R})| D^{k-1}v \quad \text{is absolutely continuous and} \quad D^k v \in L_2(\mathbf{R})\}.$$

THEOREM 5.2.1. *Suppose that we are given a data set,* $(x_1, f_1), \cdots, (x_N, f_N)$, *with* $x_1 < \cdots < x_N$. *Let* $w(x)$ *be a positive locally integrable weight function such that*

$$0 < m \leq w(x) \leq M \quad on \quad [x_1, x_N],$$

*and*

$$w(x) = 1, \quad x \notin [x_1, x_N].$$

*Further, let*

$$(5.1) \qquad (u, v)_w := \int_{\mathbf{R}} w \, D^k u \, D^k v$$

*be a semi-inner product for $H_k$. Then for $N \geq k$ there is a unique $\sigma \in H_k$ which interpolates the data and for which $(\sigma, \sigma)_w$ is a minimum. Moreover, $D^k \sigma = 0$ outside $[x_1, x_N]$.*

Proof. We first decompose $H_k$ into a direct sum $H_k^0 \oplus \mathcal{P}_{k-1}$. Specifically, let $P : H_k \to \mathcal{P}_{k-1}$ be the projector defined by interpolation at $x_1, x_2, \cdots, x_k$, so that in Lagrange form

$$(Pv)(x) := \sum_{j=1}^{k} \ell_j(x) v(x_j).$$

Clearly, the kernel of $I - P$ is just $\mathcal{P}_{k-1}$. We set $H_k^0 = Im(I - P)$. It is not difficult to see that $H_k^0$ is in fact a Hilbert space. From this it follows that an optimal interpolant $p \in H_k$ has the form $u + Pf$, where $u \in H_k^0$ interpolates the reduced data $y_i := f_i - (Pf)(x_i)$, $i = 1, \cdots, N$. Note that $y_1 = y_2 = \cdots = y_k = 0$. All $u \in H_k^0$ satisfy $u(x_j) = 0$, $1 \leq j \leq k$, so that $u$ only need to interpolate at $x_{k+1}, \cdots, x_N$. Suppose for the moment that there are representers for function evaluation in $H_k^0$; i.e., there are $K_i \in H_k^0$ such that $(K_i, u)_w = u(x_i)$ for all $u \in H_k^0$. Then it is well known (see Davis [2]) that $u$ has least norm in $H_k^0$ if and only if $u \in \text{span}\{K_i\}$. Equivalently, we wish to find $K_i \in H_k^0$ such that

$$(5.2) \qquad (K_i, v)_w = (v - Pv)|_{x=x_i} \quad i = k + 1, \cdots, N$$

for all $v \in H_k$. It seems easiest to show the existence of such $K_i$ by making use of distribution theory. Therefore, introduce the space of test functions $\mathcal{D} := \{\varphi \in C^\infty(\mathbf{R}) : \varphi \text{ has compact support}\}$. Now for any $\varphi \in \mathcal{D}$ the condition (5.2) is equivalent to

$$(5.3) \qquad < D^k(wD^k K_i), \varphi > = < \delta_{x_i}, \varphi > - \sum_{j=1}^{k} \ell_j(x_i) < \delta_{x_j}, \varphi > .$$

Of course, $\delta_x$ is the Dirac delta function and for $T$ a distribution, $< T, \varphi > := T(\varphi)$. Now such $K_i's$ can be constructed from a solution of the distributional differential equation

$$D^k(wD^k E_i) = \delta_{x_i}$$

in the following manner. This equation is known to be satisfied by a function $E_i$ such that

$$w(x)D^k E_i(x) = \Psi(x - x_i),$$

where $\Psi(x) = x^{k-1}/\{2(k-1)!\}$ for $x \geq 0$, and $\Psi(x) = -x^{k-1}/\{2(k-1)!\}$ for $x < 0$. Thus a solution is

$$(5.4) \qquad E_i(x) := \int_{x_i}^{x} \cdots \int_{x_i}^{t_2} \frac{\Psi(t_1 - x_i)}{w(t_1)} dt_1 \cdots dt_k.$$

It is not hard to see, in view of our assumptions about $w$, that $E_i \notin H_k$. However, the function

$$(5.5) \qquad F_i(x) := E_i(x) - \sum_{j=1}^{k} \ell_j(x_i) E_j(x)$$

can be shown to be in $H_k$. To see this, consider

$$\int_{\mathbf{R}} w(D^k F_i)^2 dx = \left\{ \int_{-\infty}^{x_1} + \int_{x_1}^{x_n} + \int_{x_n}^{\infty} \right\} w(D^k F_i)^2 dx.$$

Since $w(x) = 1$ for $x \notin [x_1, x_N]$, the first and last integrals vanish. For suppose $x > x_N$. Then from (5.4) and (5.5) it follows that

$$D^k F_i(x) = \Psi(x - x_i) - \sum_{j=1}^{k} \Psi(x - x_j)\ell_j(x_i) = \Psi(x - x_i) - \Psi(x - x_i),$$

since for positive arguments, $\Psi \in \mathcal{P}_{k-1}$. A similar argument shows that $D^k F_i(x) = 0$ for $x < x_1$. Consequently,

$$\int_{\mathbf{R}} w[D^k F_i]^2 dx = \int_{x_1}^{x_N} w[D^k F_i]^2 dx < \infty,$$

and so $F_i \in H_k$. A simple calculation now reveals that $F_i$ satisfies (5.3), from which it follows that

$$K_i := F_i - PF_i \in H_k^0$$

also satisfies (5.3) and $D^k K_i$ vanishes outside $[x_1, x_N]$.

We thus have shown that

$$(K_i, \varphi)_w = \varphi(x_i) - \sum_{j=1}^{k} \ell_j(x_i)\varphi(x_j), \quad i = k + 1, \cdots, N,$$

for all test functions $\varphi \in \mathcal{D}$, and this can be extended to all $v \in H_k$ by a density argument.

The problem of finding a $u \in H_k^0$ of least norm, satisfying the (hyperplane) equations

$$(K_j, u)_w = u(x_j) = y_j$$

is solved by a linear combination $\sum_{i=k+1}^{N} \alpha_i K_i(x)$ satisfying $\sum_{i=k+1}^{N} \alpha_i K_i(x_j) = y_j$. But $(K_i, K_j)_w = K_j(x_i) - (PK_j)(x_i)$ and $(PK_j)(x_i) = P(H_j - PH_j)|_{x=x_i} = (PH_j - P^2 H_j)|_{x=x_i} = 0$. It follows that the Vandermondian $[K_i(x_j)]$ is in fact a Gram matrix of $N - k$ linearly independent functions and hence positive-definite. To see the independence, suppose that there are constants $\beta_i$, not all zero, such that $K := \sum_{i=k+1}^{N} \beta_i K_i = 0$. Then $(K, v) = 0$ for all $v \in H_k^0$. Choose $v$ so that $v(x_i) = \beta_i$. Then $(K, v) = \sum \beta_i^2 \neq 0$. Hence there is a unique interpolant of minimal semi-norm, of the form $\sigma = u + Pf$, and its $k$th derivative vanishes outside $[x_1, x_N]$. As in the unweighted case, $\sigma$ is orthogonal to every interpolant in $H_k$ of zero data.

In general, computing such optimal interpolants is of course difficult. But in the case that $1/w(x)$ is piecewise polynomial, it turns out that the optimal interpolant is also piecewise polynomial and can "easily" be computed.

DEFINITION 5.2.2.   Given a knot sequence $\pi : a = x_1 < \cdots < x_N = b$, let $T_k(\pi) := \{t(x) \in C^{k-2} | t(x)$ is a spline of degree $k - 1$ and $t(x) = 0$ outside $(a, b)\}$. Note if $t \in T_k$, then $D^j t(a) = D^j t(b)$, $0 \leq j \leq k - 2$. ($C^{-1}[a, b]$ is the space of right-continuous functions on $[a, b]$).

THEOREM 5.2.2.   *Let $H_k$ be the Sobolev space above and let*

$$\pi_1 : a = x_1 < x_2 < \cdots < x_N = b \ \text{with} \ N \geq k$$

*and*

$$\pi_2 : a \leq \tau_1 < \tau_2 < \cdots < \tau_M \leq b$$

*be two partitions of $[a, b]$. Suppose that $w(x) > 0$ is a weight function such that $1/w(x)$ is piecewise polynomial of degree $m$ with respect to the partition $\pi_2$ and identically 1 outside $[a, b]$. Further, suppose that $f_1, \cdots, f_N$ are given function values to be interpolated at the points of $\pi_1$. Then the optimal interpolant $\sigma \in H_k$ which minimizes $(v, v)_w$ is such that*

(1) *$\sigma(x)$ is piecewise polynomial of degree $m + 2k - 1$ with respect to the partition $\pi_1 \cup \pi_2$,*

(2) *there is a $t \in T_k(\pi_1)$ such that $D^k \sigma(x) = t(x)/w(x)$, and*

(3) *$D^j \sigma(a^+) = D^j \sigma(b^-) = 0$, $k \leq j \leq 2k - 2$.*

*Note.* (3) is a consequence of (1) and (2) but we wish to make these "natural" end conditions explicit.

*Proof.* Suppose first that such an interpolant $\sigma \in H_k$, satisfying (1), (2) and (3), exists. Let $s \in H_k$ be any other interpolant of the given data. Then, by (2) and (3), $(s - \sigma, \sigma)_w = \int_{\mathbf{R}} D^k(s - \sigma)(x) D^k \sigma(x) w(x) dx = \int_a^b D^k(s - \sigma)(x) D^k \sigma(x) w(x) dx = \int_a^b D^k(s - \sigma)(x) t(x) dx$. Using the boundary conditions on $t(x)$, upon integrating by parts $k - 1$ times, we see that

$$(s - \sigma, \sigma)_w = (-1)^{k-1} \int_a^b D^1(s - \sigma) D^{k-1} t(x) dx.$$

But $D^{k-1}t(x)$ is piecewise constant on the partition $\pi_1$, hence

$$(s-\sigma,\sigma)_w = (-1)^{k-1}\sum_{i=1}^{N-1}\{D^{k-1}t|_{(x_i,x_{i+1})}\}\int_{x_i}^{x_{i+1}}D^1(s-\sigma)(x)dx = 0$$

since $s(x)$ and $\sigma(x)$ interpolate the same data. A standard argument shows that then, $\sigma(x)$ is optimal.

Now to show that there exists such a $\sigma \in H_k$ which satisfies (1), (2), and (3). It is well known (see, e.g., de Boor [3], p. 113) that $\dim(T_k(\pi_1)) = (N + k - 2) - 2(k - 1) = N - k$. Hence if we set

$$S_k := \{s(x) \in H_k | D^k s(x) = t(x)/w(x) \text{ for some } t \in T_k(\pi_1)\},$$

then $\dim(S_k) = \dim(T_k(\pi_1)) + k = N$. That $\sigma(x)$ satisfy (1), (2) and (3) is equivalent to $\sigma \in S_k$. Now we are asking that $\sigma(x)$ also satisfy $N$ interpolation conditions. As these are $N$ linear conditions on an $N$-dimensional space, we need only show that the corresponding homogeneous system has only the trivial solution. But if $\sigma(x) \in S_k$ is such that $\sigma(x_i) = 0$, $1 \leq i \leq N$, then by exactly the same calculation as above with $s(x) = 0$, we have $(\sigma,\sigma)_w = 0$ and so $\sigma(x) = 0$ almost everywhere. Since $\sigma \in H_k$, then $\sigma(x)$ is a polynomial of degree $k - 1$ which is zero at $N \geq k$ points and so is identically zero.

## 5.3. Computational Aspects

It is of course desirable to have an efficient means of computing such optimal splines. In [9] we describe how this may be done in case $w(x)$ is piecewise constant. We now outline a more general method that also always results in a banded system of equations. The idea of the method is perhaps best made clear by considering an important special case. To that end, suppose that $1/w(x)$ is continuous and piecewise linear. Let $\pi : a = t_1 < t_2 < \cdots < t_K = b$ be the partition $\pi_1 \cup \pi_2$ and suppose that $v_i$ is the value of $1/w(x)$ at $t_i$, $1 \leq i \leq K$. Further, take $k = 2$, so that we wish to compute the optimal interpolant that minimizes $\int_{\mathbf{R}}(D^2\sigma(x))^2 w(x)dx$. By Theorem 5.2.2, $\sigma$ will be piecewise polynomial of degree $3 + 1 = 4$ on the partition $\pi$. Moreover, as $1/w(x)$ is assumed to be continuous, $D^2\sigma \in C[a,b]$ so that $\sigma \in C^2[a,b]$. Now the dimension of the space of $C^2$ piecewise quartics on the partition $\pi$, is $5(K - 1) - 3(K - 2) = 2K + 1$, and a basis for this space may easily seen to be the first $2K + 1$ quartic $B$-splines based on the knot sequence

$$t_{-2},t_{-1},t_0,t_1,t_1,t_2,t_2,\cdots,t_K,t_K;$$

i.e., each knot of $\pi$ is repeated once and three additional knots are adjoined to the left end. Let us refer to this basis as $\{B_i\}_1^{2K+1}$. Clearly, our optimal interpolant $\sigma = \sum_{i=1}^{2K+1}\alpha_i B_i$ for some scalars $\alpha_i$, and we now proceed to find $2K + 1$ linear equations that these coefficients must satisfy. First note that since $D^2\sigma(x) = t(x)/w(x)$ for some $w(x) \in T_2(\pi_1)$, it follows that on each interval

$[t_i, t_{i+1}]$ the polynomial with which $D^2\sigma(x)$ agrees must be divisible by the corresponding linear portion of $1/w(x)$. This is easily seen to be equivalent to the condition:

$$0 = v_{i+1}(v_i + v_{i+1})D^2\sigma(t_i) - 4v_i v_{i+1}D^2\sigma((t_i + t_{i+1})/2) + v_i(v_i + v_{i+1})D^2\sigma(t_{i+1}).$$

This gives $K - 1$ conditions. Further, we have $N$ interpolation conditions at the $N$ points of $\pi_1$, and if $t_i \in \pi$ is *not* one of the $x_i$, by (2) of Theorem 5.2.2, $w(x)D^2\sigma(x)$ does *not* have a knot at $t_i$; i.e., is smooth there. Since $w(x)D^2\sigma(x)$ is piecewise linear, this is equivalent to

$$D(wD^2\sigma)(t_i^+) = D(wD^2\sigma)(t_i^-).$$

This gives $K - N$ additional conditions which together with the two natural end conditions, (3) of Theorem 5.2.2, gives a total of

$$(K - 1) + N + (K - N) + 2 = 2K + 1.$$

The bandedness of the system results from the use of the $B$-spline basis.

## 5.4.  Limits of Weighted Splines

We next show that optimal interpolants based on general weight functions may be computed by taking the limit of optimal interpolants based on simpler weight functions.

THEOREM 5.4.1.  *Given a data set* $(x_1, f_1), \cdots, (x_N, f_N)$ *with* $x_1 < \cdots < x_N$ *and* $N \geq k$, *suppose that* $w(x)$ *is a weight function satisfying the conditions of Theorem 5.2.1. Suppose further that* $w_n(x)$, $n = 1, 2, \cdots$, *is a sequence of weight functions, also satisfying the conditions of Theorem 5.2.1, such that* $0 < m \leq w_n(x)$ *on* $[x_1, x_N]$ *and that* $\lim_{n \to \infty}\{\sup_{x \in [x_1, x_N]} |w(x) - w_n(x)|\} = 0$. *If* $\sigma(x)$ *is that element of* $H_k$ *which interpolates the given data* $\{(x_i, f_i) : 1 \leq i \leq N\}$ *for which* $|\sigma|_w$ *is a minimum and* $s_n$ *the corresponding minimizer with weight* $w_n(x)$, *then*

$$s_n \to \sigma \ \text{uniformly on} \ [x_1, x_N].$$

*Proof.* First, given $\epsilon > 0$, choose $M$ so that $n > M \Rightarrow |w(x) - w_n(x)| \leq \epsilon m$ on $[x_1, x_N]$. Then, for $n > M$,

$$|(s_n, s_n)_w - (s_n, s_n)_{w_n}| = |\int_{x_1}^{x_N} (D^k s_n(x))^2 (w(x) - w_n(x))dx|$$

$$\leq \int_{x_1}^{x_N} (D^k s_n(x))^2 |w(x) - w_n(x)|dx$$

$$\leq \int_{x_1}^{x_N} (D^k s_n(x))^2 \epsilon m \, dx$$

$$\leq \epsilon \int_{x_1}^{x_N} (D^k s_n(x))^2 w_n(x)dx \quad (\text{as} \quad m \leq w_n(x))$$

$$\leq \epsilon \int_{x_1}^{x_N} (D^k \sigma(x))^2 w_n(x)dx \quad (\text{by the minimality of } s_n).$$

But

$$\int_{x_1}^{x_N} (D^k\sigma(x))^2 w_n(x)dx \to \int_{x_1}^{x_N} (D^k\sigma(x))^2 w(x)dx,$$

and so we see that

$$|(s_n, s_n)_w - (s_n, s_n)_{w_n}| \to 0.$$

Now, by the minimality of $s_n$, $(s_n, s_n)_{w_n} \le (\sigma, \sigma)_{w_n}$, and by the minimality of $\sigma$, $(\sigma, \sigma)_w \le (s_n, s_n)_w$. But also, $(\sigma, \sigma)_{w_n} \to (\sigma, \sigma)_w$ and thus we must have $(s_n, s_n)_w \to (\sigma, \sigma)_w$. Further, the optimality of $\sigma$ implies that $(\sigma, \sigma - s_n)_w = 0$ and so

$$(s_n, s_n)_w = (\sigma - (\sigma - s_n), \sigma - (\sigma - s_n))_w = (\sigma, \sigma)_w + (\sigma - s_n, \sigma - s_n)_w,$$

and we see that

$$(\sigma - s_n, \sigma - s_n)_w = (s_n, s_n)_w - (\sigma, \sigma)_w \to 0.$$

Therefore, as $\frac{w(x)}{m} \ge 1$ on $[x_1, x_N]$,

$$(\sigma - s_n, \sigma - s_n)_1 \le \frac{1}{m}(\sigma - s_n, \sigma - s_n)_w \to 0.$$

(Here $(\cdot, \cdot)_1$ denotes the semi-inner product with weight function 1.) This shows convergence in an $L_2$ sense. To show pointwise convergence, consider

$$E_t(x) := (-1)^{k-1}\{|x - t|^{2k-1} - \sum_{j=1}^{k} \ell_j(t)|x - x_i|^{2k-1}\}/(2(2k - 1)!)$$

where the $\ell_j$ are the Lagrange interpolating polynomials for the $k$ points $x_1, \cdots, x_{k-1}$ and $x_N$. It is easily seen that, in fact, $E_t \in H_{2k-1} \subset H_k$ for any $t \in \mathbf{R}$ and that $E_t^{(j)}(x) = E_t^{(j)}(x) = 0$, $k \le j \le 2k - 2$, outside $[x_1, x_N]$. Integration by parts then reveals that

$$(h, E_t)_1 = \int_{x_1}^{x_N} D^k h(x) D^k E_t(x)dx = h(t)$$

for $t \in [x_1, x_N]$, and $h \in H_k$ which is zero at these $k$ points. Thus, as $\sigma$ and $s_n$ are interpolants of the same data,

$$|(\sigma - s_n)(t)|^2 = |(\sigma - s_n, E_t)_1|^2 \le (\sigma - s_n, \sigma - s_n)_1(E_t, E_t)_1 \to 0.$$

The convergence is uniform on $[x_1, x_N]$ for $(E_t, E_t)_1$ is evidently uniformly bounded on $[x_1, x_N]$.

A natural question arising at this point concerns the choice of weight function. We will now show that if the data originates from a function $f$ which is spline-like in a sense defined below, and if its $k$th derivative is known, then there is a weight function $w$ such that the optimal interpolant $\sigma = f$. This $w$

can be approximated by piecewise polynomial weight functions $w_n$ converging uniformly to $w$. By Theorem 5.2.2, the corresponding optimal interpolating splines are also piecewise polynomial, which, by Theorem 5.4.1, converge to $f$.

DEFINITION 5.4.1.    Suppose $\pi : x_1 < \cdots < x_N$ with $N \geq k$, and $f \in C^{2k-2}[x_1, x_N]$. Then we say that $f$ is spline-like of degree $2k - 1$ on this partition if and only if $D^j f(x_1) = D^j f(x_N) = 0$, $k \leq j \leq 2k - 2$, and there exists a $t \in T_k(\pi)$ such that $w(x) := t(x)/D^k f(x) \in C[x_1, x_N]$ and $w(x) > 0$ on $[x_1, x_N]$. It follows that $t(x)/D^k f(x)$ can be extended to a weight function $w$ satisfying the conditions of Theorem 5.4.1 by defining

$$w(x) := \begin{cases} t(x)/D^k f(x), & x \in [x_1, x_N], \\ 1, & x \notin [x_1, x_N]. \end{cases}$$

THEOREM 5.4.2.   *Suppose that $f(x)$ is spline-like of degree $2k - 1$ with respect to the partition $x_1 < \cdots < x_N$. Then the optimal interpolant of the data $(x_i, f_i)$, $1 \leq i \leq N$, from $H_k$, with respect to the weight function $w(x)$ defined above, is $f(x)$ itself.*

*Proof.* We extend $f$ by its Taylor polynomials of degree $k - 1$ at $x_1$ and $x_N$ (and continue to refer to it as $f$) to a function $f \in H_k$, and recall that $f$ is optimal if and only if it is orthogonal in the weighted semi-inner product to every interpolant of zero data at the given points. Let $z \in H_k$ be such an interpolant. Then

$$\int_{\mathbf{R}} w(x) D^k f(x) D^k z(x) dx = \int_{x_1}^{x_N} w(x) D^k f(x) D^k z(x) dx = \int_{x_1}^{x_N} t(x) D^k z(x) dx.$$

Now integrating by parts $k - 1$ times and using the boundary conditions on $t(x)$, we see that

$$(f, z)_w = (-1)^{k-1} \int_{x_1}^{x_N} D^1 z(x) D^{k-1} t(x) dx.$$

But $D^{k-1} t(x)$ is piecewise constant on the partition, hence we obtain

$$(-1)^{k-1} \sum_{i=1}^{N-1} \{ D^{k-1} t(x)|_{(x_i, x_{i+1})} \} \int_{x_i}^{x_{i+1}} D^1 z(x) dx,$$

which is zero as $z(x)$ is zero at each of the $x_i$.

We note that similar results hold for clamped splines.

## 5.5.   Examples

We apply the above to a data set which leads to an unreasonably oscillatory natural cubic spline interpolant. In the first illustration, we choose a continuous piecewise linear reciprocal weight function and construct the corresponding $C^2$ quartic spline. The first step is to sketch an interpolant. This "free-hand"

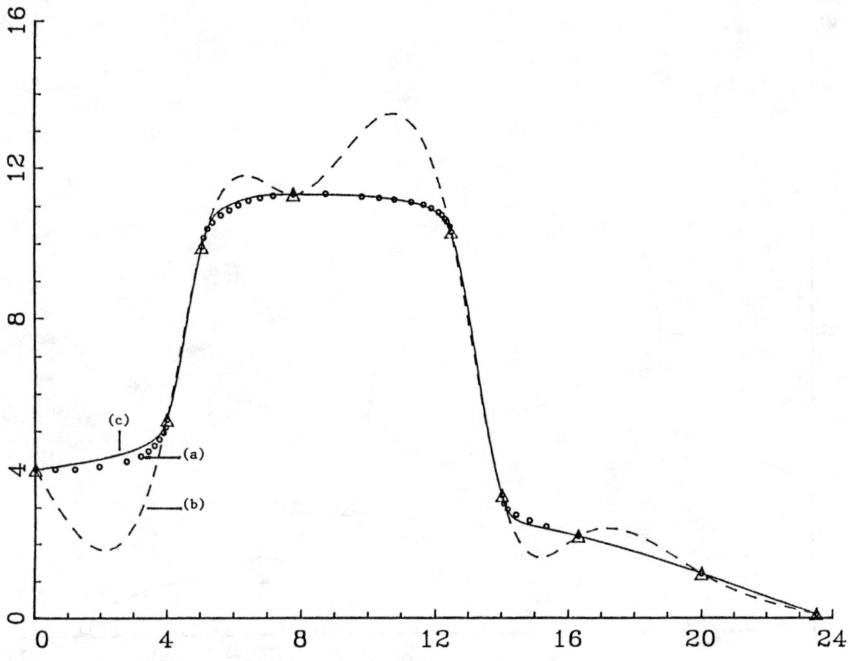

FIG. 5.1. (a) *Sketched curve,* (b) *cubic spline, and* (c) *quartic spline.*

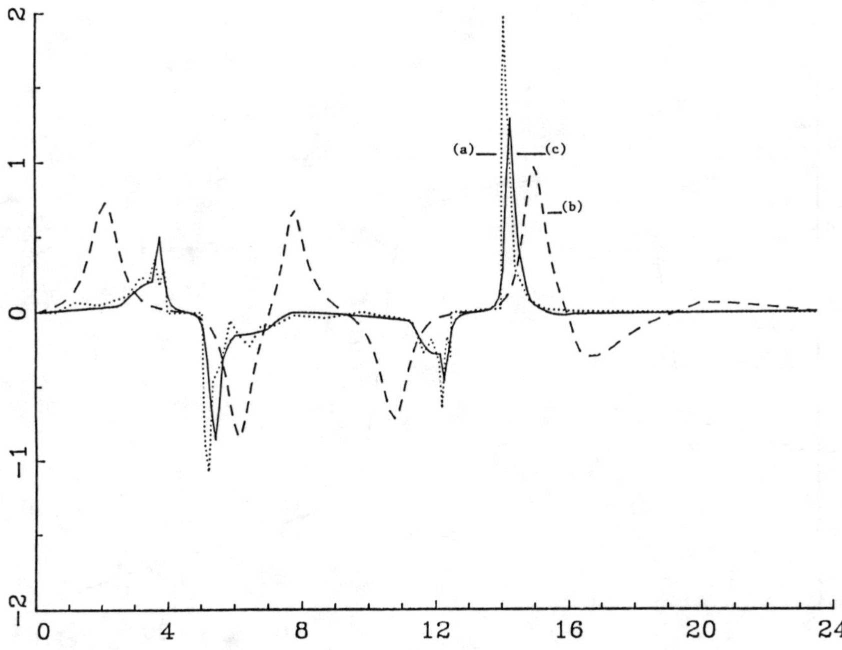

FIG. 5.2. *Curvature for* (a) *sketched curve,* (b) *cubic spline, and* (c) *quartic spline.*

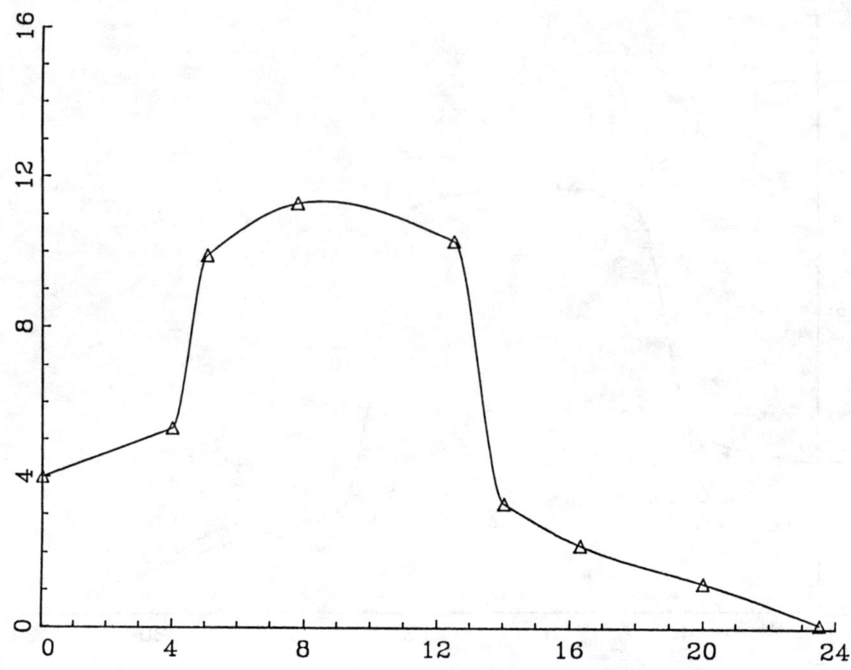

FIG. 5.3. *Cubic spline with piecewise constant weights.*

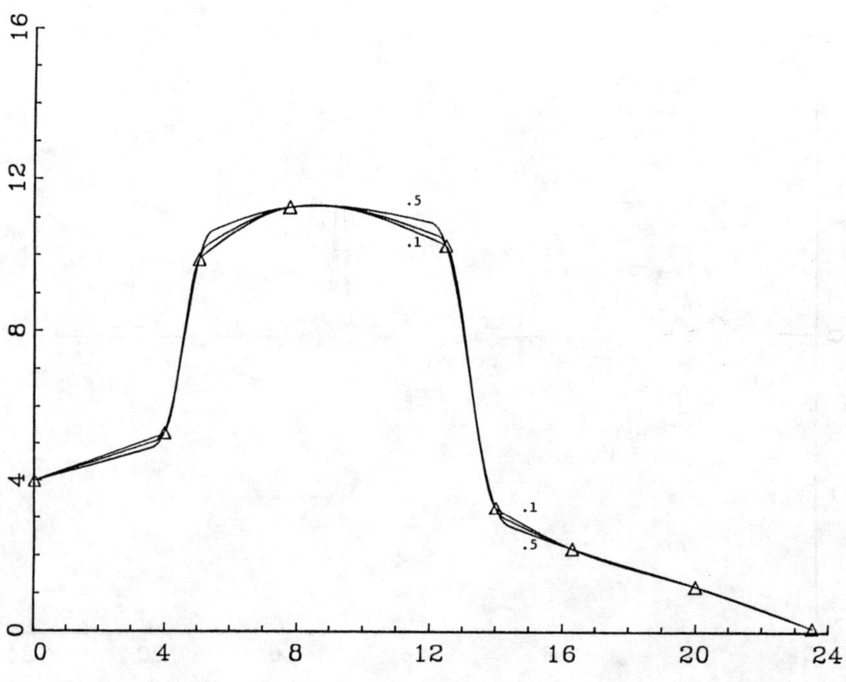



FIG. 5.4. *Quartic spline with smoothed weights:* $\epsilon = 0.1,\ 0.3,\ and\ 0.5.$

data supplies an approximation to the second derivative via divided differences. It seems essential to sample this sketch at high density. It is verified that the sketched curve is indeed spline-like with respect to the given abscissas, and an appropriate continuous linear spline $t(x)$ is created to conform with the specifications in Definition 5.4.1. We now have a weight function whose reciprocal is approximated by a continuous linear spline $v(x)$ whose knots form the partition $\pi_2$ of Theorem 5.2.2. The $t(x)$ and $v(x)$ are conveniently constructed in an interactive manner. The final step is to compute the quartic spline which interpolates the given data and approximates the sketch. One expects the spline to be smoother than the free-hand sketch. Figure 5.1 shows the natural cubic spline interpolant, the sketch, and the weighted quartic natural spline interpolant. In Fig. 5.2 we show the (signed) curvature, and note that as expected the cubic spline has a much smaller maximum magnitude of the curvature; for this data this is not appropriate. The quartic spline follows the curvature of the sketch quite closely, and seems to be somewhat less extreme.

The second example uses the same data, interpolating it with a weighted $C^1$ cubic spline using a piecewise constant weight function as in [8]. The reciprocal of this weight function is also piecewise constant. The interpolant, shown in Fig. 5.3, is quite angular; some discussion of this will be found in [9]. We now smooth the reciprocal weight by convolution with a unit pulse of small width $\epsilon$. The result is a continuous linear spline which cuts the corners off the piecewise constant function and whose knots form a partition $\pi_2$ as before. Associated with this weight is a $C^2$ quartic spline of more natural form, shown in Fig. 5.4 for $\epsilon = 0.1, 0.3$, and 0.5. There is still some angularity which could be reduced by cutting more corners on the piecewise linear reciprocal weight function.

## Acknowledgments

The research of these authors was supported in part by the National Sciences and Engineering Research Council of Canada under grants OGP0008389 and OGP0008621, respectively.

## References

[1]  L. Bos and K. Salkauskas, *Limits of weighted splines based on piecewise constant weight functions*, Rocky Mountain J. Math, to appear.

[2]  P. J. Davis, *Interpolation and Approximation*, Blaisdell, New York, 1963.

[3]  C. de Boor, *A Practical Guide to Splines*, Springer-Verlag, New York, 1978.

[4]  T. A. Foley, *Weighted bicubic spline interpolation to rapidly varying data*, ACM Trans. Graphics, 6 (1987), pp. 1–18.

[5]  ——, *Local control of interval tension using weighted splines*, Comput. Aided Geom. Des., 3 (1986), pp. 281–294.

[6]  ——, *Interpolation with interval and point tension controls using cubic omega splines*, ACM Trans. Math. Software, 13 (1987), pp. 68–96.

[7]  J. Meinguet, *Multivariate Interpolation at Arbitrary Points Made Simple*, Rapport No. 118, Seminaire de mathématique appliquée et mechanique, Institut de Mathématique Pure et Appliquée, Université Catholique de Louvain, Louvain-

La-Neuve, 1978.

[8]  K. Salkauskas, $C^1$ *splines for interpolation of rapidly varying data*, Rocky Mountain J. Math., 14 (1984), pp. 239–250.

[9]  K. Salkauskas and L. P. Bos, *Weighted splines as optimal interpolants*, Rocky Mountain J. Math., to appear.

# Algorithms for Geometric Spline Curves

Matthias Eck

## 6.1. Introduction

B-spline curves and surfaces are extensively used in most of existing CAD/CAM systems. The basic properties and algorithms of B-spline curves have been exhaustively investigated in the past and are summarized in the first part of §6.2. As a generalization of B-spline curves the geometric continuous B-spline curves have become more and more widespread in the last five years. These curves are constructed by generalization of the continuity conditions between the piecewise polynomial segments using some simple results of differential geometry. This kind of curve is introduced in the second part of §6.2. Unfortunately only a few algorithms concerning geometric B-spline curves of degree 3 have been published until now, although geometric B-spline curves up to degree 5 have been explicitly known.

   Therefore, the aim of this paper is to show how to modify some algorithms in the case of geometric B-spline curves which are well known in the usual $C^{n-1}$-continuity case, i.e., the insertion of a new knot, recurrence formulas, and de Boor-like algorithms. Additionally, a recursive algorithm for the derivatives of a curve is deduced.

## 6.2. Fundamentals

### 6.2.1. Usual B-Spline Curves and their Algorithms.
A usual piecewise $C^{n-1}$-continuous nonrational B-spline curve $\mathbf{X}(u) \in I\!\!R^s$ of degree $n$ can be represented by

$$\mathbf{X}(u) = \sum_{i=0}^{N} \mathbf{d}_i N_i^n(u) , \quad u \in [u_n, u_{N+1}]$$

if a monotonic increasing knot sequence $\mathcal{U} = \{u_0, \cdots, u_{N+n+1} | u_i \in I\!\!R\}$ and control points $\mathbf{d}_i \in I\!\!R^s$ are given. The basis functions $N_i^n(u) \in I\!\!R$ are called the $i$th (normalized) B-spline functions of degree $n$ and are defined by the well-known Mansfield/de Boor/Cox recursion

$$N_l^n(u) \;=\; \alpha_{l,n}(u)\cdot N_l^{n-1}(u) + (1 - \alpha_{l+1,n}(u))\cdot N_{l+1}^{n-1}(u)\,,$$

(6.1)
$$\alpha_{l,n}(u) \;=\; \frac{u - u_l}{u_{l+n} - u_l}\,, \qquad N_l^0(u) = \begin{cases} 1\,, & u \in [u_l, u_{l+1}[ \\ 0\,, & \text{else} \end{cases}.$$

Two properties of the $N_l^n(u)$ should be mentioned here:

1. *Partition of unity* : $\sum_{i=0}^N N_i^n(u) \equiv 1$ , $u \in [u_n, u_{N+1}]$

2. *Positivity* : $N_i^n(u) \begin{cases} > 0\,, & u \in ]u_i, u_{i+n+1}[ \\ = 0\,, & \text{else} \end{cases}$.

Recursion (6.1) can be used to derive a recursive algorithm for determining the value of $\mathbf{X}(u) = \mathbf{d}_l^n$ from the control points (*de Boor algorithm*):

$$\mathbf{d}_r^j = \alpha_{r,n-j+1}(u)\cdot \mathbf{d}_r^{j-1} + (1 - \alpha_{r,n-j+1}(u))\cdot \mathbf{d}_{r-1}^{j-1}\,; \quad \mathbf{d}_r^0 = \mathbf{d}_r$$

$$u \in [u_l, u_{l+1}[\,; \quad (r = l - n + j, \cdots, l)\,; \quad (j = 1, \cdots, n)\,.$$

A second important recursion formula can be formulated using the idea of *inserting a new knot* $\hat{u}$ into the given knot sequence. $\hat{N}_l^n(u)$ denotes the B-spline function over the new knot sequence, then it yields

$$\hat{N}_l^n(u) \;=\; \hat{\alpha}_{l,n}(\hat{u})\cdot \hat{N}_l^n(u) + (1 - \hat{\alpha}_{l+1,n}(\hat{u}))\cdot \hat{N}_{l+1}^n(u)\,,$$

(6.2)
$$\hat{\alpha}_{l,n}(\hat{u}) \;=\; \frac{\hat{u} - u_l}{u_{l+n} - u_l}\,.$$

Now, taking advantage from the fact that $N_j^n(u_l) = \begin{cases} 1\,, & j = l \\ 0\,, & else \end{cases}$ if $u_l$ has multiplicity $n$, (6.2) can be applied for calculating $\mathbf{X}(\hat{u}) = \hat{\mathbf{d}}_l^n$ recursively, as follows

$$\hat{\mathbf{d}}_r^j = \hat{\alpha}_{r,n-j+1}(\hat{u})\cdot \hat{\mathbf{d}}_r^{j-1} + (1 - \hat{\alpha}_{r,n-j+1}(\hat{u}))\cdot \hat{\mathbf{d}}_{r-1}^{j-1}\,; \quad \hat{\mathbf{d}}_r^0 = \mathbf{d}_r$$

$$\hat{u} \in [u_l, u_{l+1}[\,; \quad (r = l - n + j, \cdots, l)\,; \quad (j = 1, \cdots, n)\,.$$

For references and historical details the reader is referred to [9],[13].

**6.2.2. Geometric B-Spline Curves.** A method for constructing the polygons of the composite Bézier spline curve, i.e., a piecewise polynomial curve where each piece $\mathbf{X}_I(t)$ , $t \in [0, 1]$ is represented by

$$\mathbf{X}_I(t) = \sum_{k=0}^n \mathbf{b}_{nI+k} B_k^n(t)\,, \quad B_k^n(t) = \binom{n}{k} t^k (1 - t)^{n-k}\,, \quad \mathbf{b}_{nI+k} \in I\!R^s$$

from the B-spline polygon was developed by Sablonniére [18] by embedding the $C^{n-1}$-continuity construction of Bézier spline curves into a B-spline construction (Fig. 6.1).

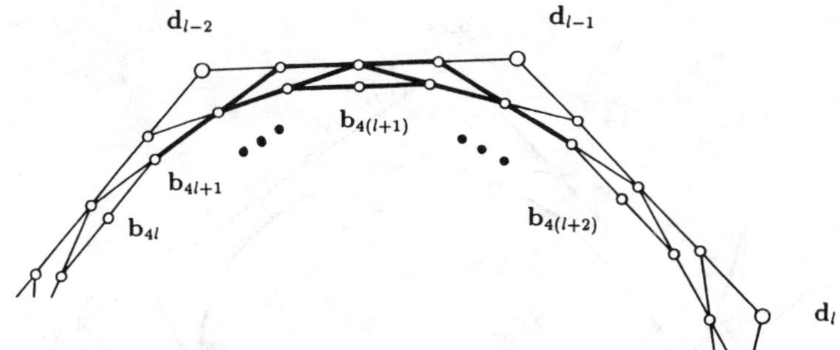



FIG. 6.1.  *Connection between Bézier and B-spline polygons, $n = 4$.*

Now, if the segments of a composite Bézier spline curve are not joining together with $C^{n-1}$-continuity but with continuity of the Frenet frame and the first $n-2$ curvatures $\kappa_i$ (called $GC^{n-1}$-continuity or $G^{n-1}$-continuity or Frenet frame-continuity, see, for example, [6]), the $C^{n-1}$-continuity construction for Bézier spline curves has to be modified at the knot $u_i$ by using additional tension or design parameters denoted by small greek letters ($n = 3$: $\gamma_i$; $n = 4$: $\gamma_i, \delta_i, \epsilon_i$; $n = 5$: $\gamma_i, \delta_i, \epsilon_i, \rho_i, \sigma_i, \tau_i$). This idea originates from Boehm [2],[3] and was enlarged to general degree by Lasser and Eck [15]. To embed this $GC^{n-1}$-construction in a B-spline construction some auxiliary parameters denoted by capital greek letters are necessary ($n = 4$: $\Gamma_i, \overline{\Gamma}_i$; $n = 5$: $\Theta_i, \overline{\Theta}_i, \Lambda_i, \overline{\Lambda}_i, \Phi_i, \overline{\Phi}_i, \Psi_i, \overline{\Psi}_i, \Omega_i, \overline{\Omega}_i$). Due to the *Theorem of Menelao* these auxiliary parameters are in special dependencies to the given design parameters which can be found in Appendix A. In Figs. 6.2 to 6.4 the treated cases $n = 3$ to $n = 5$ are described. It is obvious how to extend the concept to arbitrary degree $n$. The relationship between the Bézier points and the de Boor points can be expressed in the following way using real factors $b_{nl+i,l-j}$ which are a simple consequence from Figs. 6.2 to 6.4:

$$(6.3) \quad \mathbf{b}_{nl+i} = \sum_{j=1}^{n} b_{nl+i,l-j} \cdot \mathbf{d}_{l-j} , \quad (i = 0, \cdots, n-1) , \quad b_{nl+i,l-j} \in I\!R$$

where

1. $b_{nl+i,l-n} = 0 , \quad (i = 1, \cdots, n-1)$;

2. $b_{nl+i,l-2} = 1 - b_{nl+i,l-1} - \sum_{j=3}^{n} b_{nl+i,l-j} , \quad (i = 0, \cdots, n-1) ; \quad n \geq 3$.
The remaining factors $b_{nl+i,l-j}$ are explicitly written in Appendix B.

Here and in the following, we assume $C^1$-continuity which can be achieved by a linear parameter transformation from $GC^1$-continuity (or *tangent continuity*) [3].

Meanwhile Goodman presented an alternative construction for generating the Bézier points of a $GC^{n-1}$-continuous piecewise curve of general degree $n$ [11], but this is not treated in this paper.

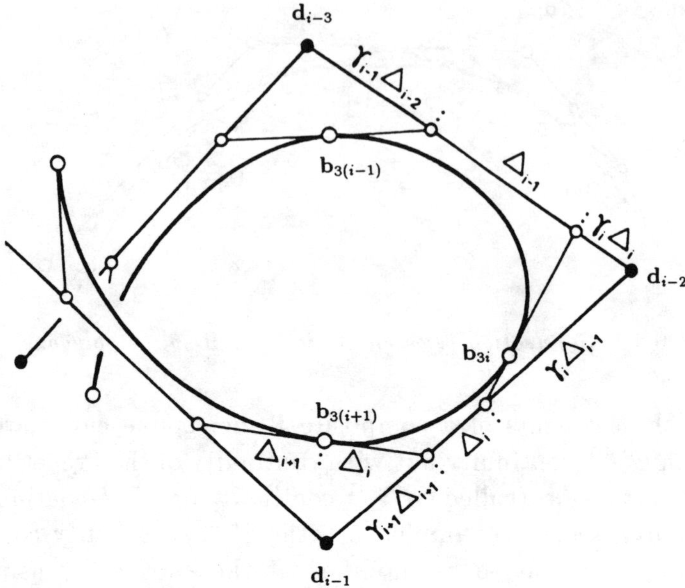

FIG. 6.2.  *Construction of a $GC^2$-continuous cubic.*

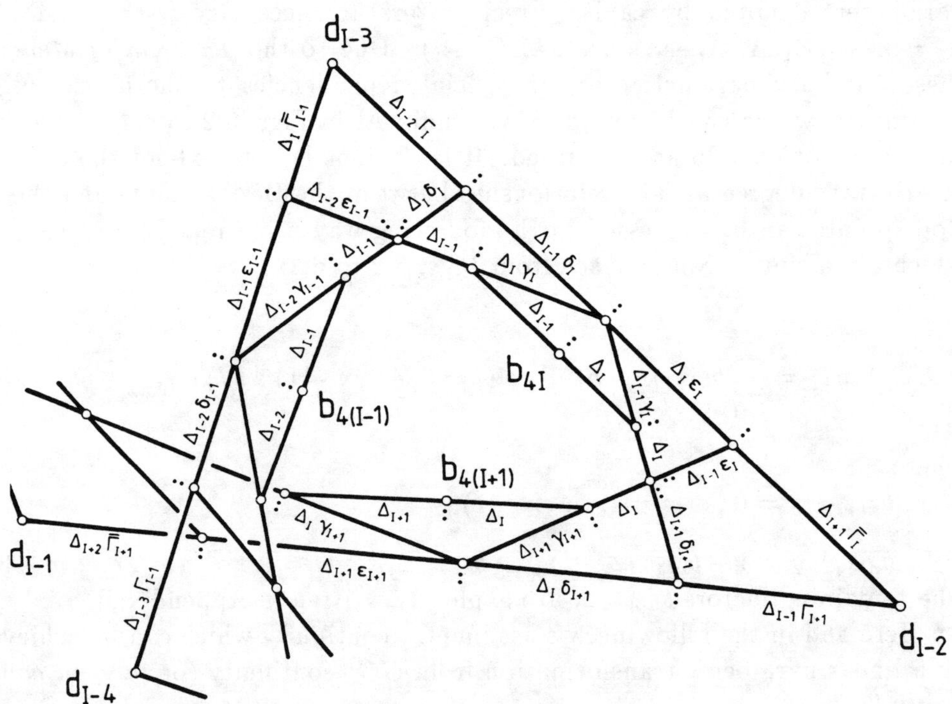

FIG. 6.3.  *Construction of a $GC^3$-continuous quartic.*

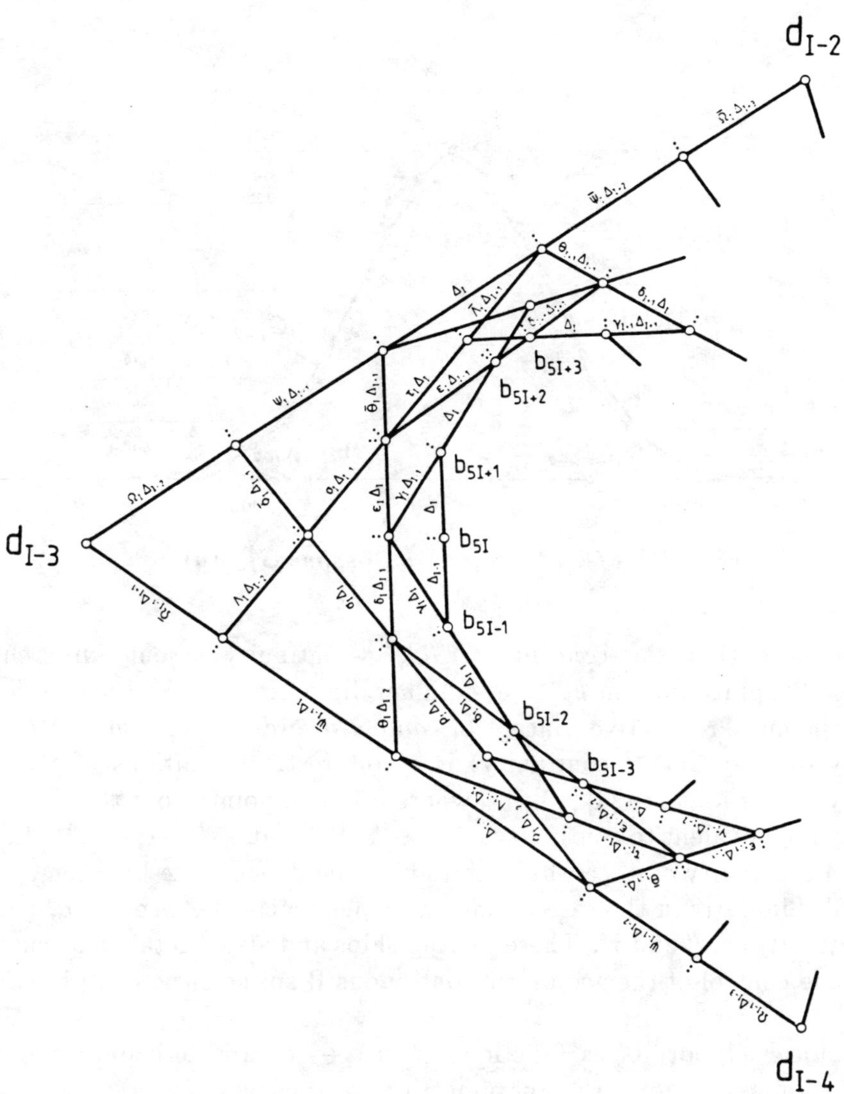

FIG. 6.4. *Construction of a $GC^4$-continuous quintic.*

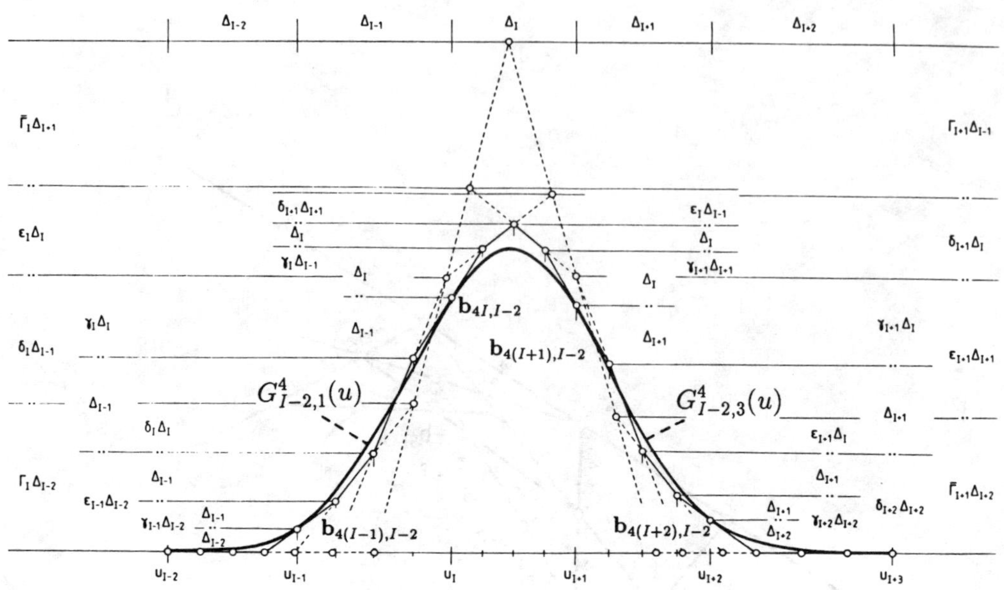

FIG. 6.5.  *Quartic geometric B-spline $G^4_{I-2}(u)$.*

Please note that the concept of $GC^{n-1}$-continuity is only differential geometrically appropriate in $I\!R^{n-1}$ and higher dimensions. For curves in $I\!R^3$ or even $I\!R^2$ the more restrictive concept of *contact of order* $n-1$ (called $VC^{n-1}$-continuity or also $GC^{n-1}$-continuity) is of interest. Definitions of $VC^{n-1}$-continuity can be found in [4],[10],[17] where it is also pointed out that $VC^{n-1}$-continuity is equivalent to continuity of $\kappa',\cdots,\kappa^{(n-3)}$ and $\tau',\cdots,\tau^{(n-4)}$. In the case $n=4$ continuity of $\kappa'$ can be achieved by one dependence between $\delta_i$ and $\epsilon_i$ [14],[16]. Similarly in the case $n=5$ two dependencies between $\rho_i,\sigma_i$ and $\tau_i$ fulfill continuity of $\kappa''$ and $\tau'$. These relationships and also a detailed discussion of the shape control of the geometric continuous B-spline curves can be found in [8],[14].

Also, local support basis functions $G^n_i(u) \in I\!R$ are derivable in Bézier representation as segmented nonparametric Bézier curves $G^n_{i,r}(u)$ :

$$G^n_{i,r}(u) = \sum_{j=0}^{n} b_{n(i+r)+j,i} B^n_j \left( \frac{u - u_{i+r}}{\triangle_{i+r}} \right) \, , \quad u \in [u_{i+r}, u_{i+r+1}] \, , \quad (r = 0, \cdots, n).$$

(6.4)

The Bézier ordinates $b_{n(i+r)+j,i}$, which are given in (6.3), have to be carried up over the abscissae $u_{i+r} + \frac{j}{n}\triangle_{i+r}$, as in Fig. 6.5 for $n = 4$. Because our construction is assumed to be $C^1$-continuous, it follows that $G^s_i(u) = N^s_i(u)$ , $(s = 1,2)$.

Please note first that also the $G^n_i(u)$ build a partition of unity which immediately follows from the construction and from the fact that the Bernstein polynomials build a partition of unity and second that in the case $n = 3$

the $G_i^n(u)$ are always positive. In the cases $n = 4$ and $n = 5$ some simple inequalities are necessary to fulfill this property [8].

## 6.3.  Insertion

The first algorithm which has been adapted to geometric continuous B-spline curves is the often used procedure of inserting a new knot into B-spline curves. This is a very powerful tool for manipulating B-spline curves because it does not change the shape of the curve and therefore gives the designer more control over the shape.

In the case of geometric splines the technique can be expressed as follows.

Assume the original knot sequence $\mathcal{U}$; a new knot $\hat{u} \in ]u_l, u_{l+1}[$; and the original design parameter vector $\mathcal{P}$ in which all design parameters of the interior knots are listed to be given. For example in the case $n = 4$ this design parameter vector $\mathcal{P}$ looks like $\{\gamma_5, \cdots, \gamma_N, \delta_5, \cdots, \delta_N, \epsilon_5, \cdots, \epsilon_N\}$. At first a refined knot sequence $\hat{\mathcal{U}}$ given by

$$\begin{aligned}
\hat{u}_i &= u_i, & i \le l \\
\hat{u}_{l+1} &= \hat{u} \\
\hat{u}_i &= u_{i-1}, & i \ge l+2
\end{aligned}$$

and a new design parameter vector $\hat{\mathcal{P}}$ given by

$$\begin{aligned}
(6.5) \quad \hat{\gamma}_i &= \gamma_i, & \hat{\delta}_i &= \delta_i, & \hat{\epsilon}_i &= \epsilon_i, & \cdots & , i \le l-1 \\
(6.6) \quad \hat{\gamma}_l &= \hat{\gamma}^1, & \hat{\delta}_l &= \hat{\delta}^1, & \hat{\epsilon}_l &= \hat{\epsilon}^1, & \cdots & \\
(6.7) \quad \hat{\gamma}_{l+1} &= 1, & \hat{\delta}_{l+1} &= 1, & \hat{\epsilon}_{l+1} &= 1, & \cdots & \\
(6.8) \quad \hat{\gamma}_{l+2} &= \hat{\gamma}^2, & \hat{\delta}_{l+2} &= \hat{\delta}^2, & \hat{\epsilon}_{l+2} &= \hat{\epsilon}^2, & \cdots & \\
(6.9) \quad \hat{\gamma}_i &= \gamma_{i-1}, & \hat{\delta}_i &= \delta_{i-1}, & \hat{\epsilon}_i &= \epsilon_{i-1}, & \cdots & , i \ge l+3
\end{aligned}$$

are constructed.  Equations (6.5) and (6.9) immediately follow from the property of local support of the $G_i^n(u)$. Equation (6.7) is obvious because at $\hat{u}$ the resulting curve is $C^n$-continuous which is only possible if all design parameters are equal to 1. The values of (6.6) and (6.8) denoted by superscripts 1 and 2 are unknown and have to be determined in dependence to $\hat{u}$.

Then a spline curve $\mathbf{X}(u) = \sum_{i=0}^{N} \mathbf{d}_i G_i^n(u)$ should be written in terms of the new local basis function $\hat{G}_i^n(u)$, built by the same construction as $G_i^n(u)$ only concerning $\hat{\mathcal{U}}$ and $\hat{\mathcal{P}}$, as $\mathbf{X}(u) = \sum_{i=0}^{N+1} \hat{\mathbf{d}}_i \hat{G}_i^n(u)$ where the new de Boor points should be calculated by

$$(6.10) \quad \hat{\mathbf{d}}_i = \begin{cases} \mathbf{d}_i, & i \le l-n \\ \hat{\alpha}_{i,n}(\hat{u}) \cdot \mathbf{d}_i + (1 - \hat{\alpha}_{i,n}(\hat{u})) \cdot \mathbf{d}_{i-1}, & l-n+1 \le i \le l \\ \mathbf{d}_{i-1}, & i \ge l+1 \end{cases}.$$

That means the insertion procedure is completely determined if the dependencies of the unknowns $\hat{\gamma}^1, \hat{\gamma}^2, \hat{\delta}^1, \hat{\delta}^2, \cdots$ and $\hat{\alpha}_{l-n+1,n}, \cdots, \hat{\alpha}_{l,n}$ on $\hat{u}$

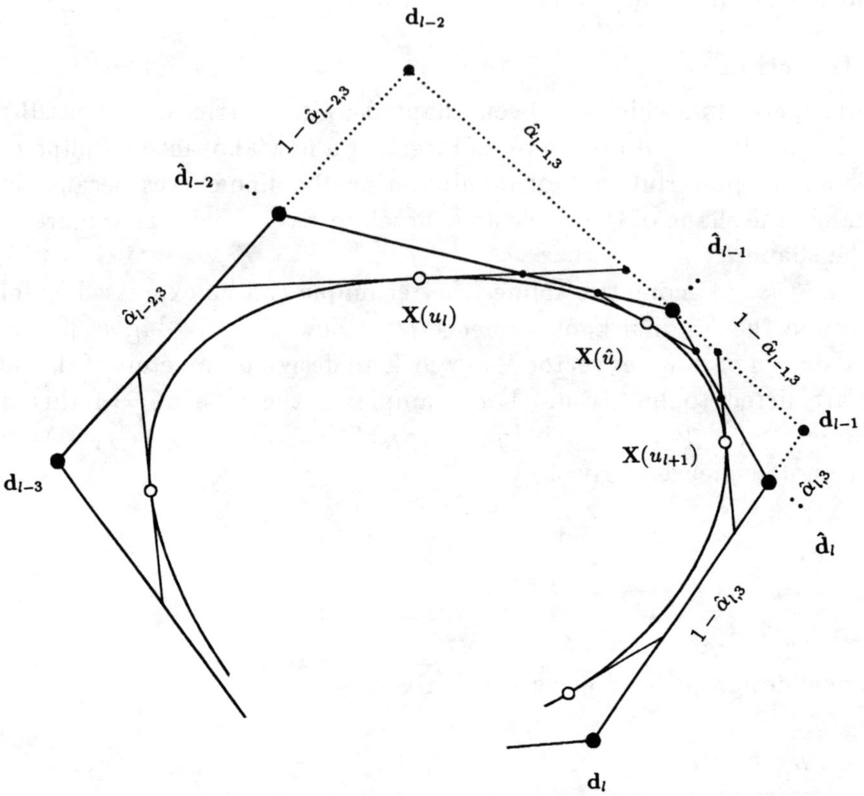



FIG. 6.6.  *Insertion of a new knot in the cubic case.*

are known.  To get such equations, we use a geometric approach to the problem of inserting knots which was developed by Boehm in the cubic case [2]. Alternatively, Dierckx and Tytgat [5] described a more direct way to get an inserting procedure for cubic Beta-spline curves, a representation of $GC^2$-continuous B-spline curves based on monoms [1].

As already said in §6.2, the geometric B-splines can be interpreted as a construction for a composite Bézier spline curve.  On the other hand it is well known that the Bézier points of a subdivision at $\hat{u}$ of a segment defined over $[u_l, u_{l+1}]$ can be calculated by the algorithm of de Casteljau.  Now, the embedding of the refined Bézier spline curve into a B-spline construction causes a change of the original de Boor polygon. See Fig. 6.6 for the example $n = 3$ [2] where the two control points $d_{l-1}$ and $d_{l-2}$ are to be replaced by $\hat{d}_{l-2}$, $\hat{d}_{l-1}$ and $\hat{d}_l$.  These points are evaluated by applying again the *Theorem of Menelaos*. The extension of this procedure of embedding the de Casteljau algorithm to the treated cases $n = 4$ and $n = 5$ is obviously also possible.

Summarizing the new design parameters in the cubic and quartic case,

(6.11)

$$\hat\gamma^1 = \frac{\gamma_l \Delta_l}{\hat\Delta^1 + \dfrac{\gamma_l \Delta_{l-1} + \hat\Delta^1}{\Delta_{l-1} + \hat\Delta^1}\,\hat\Delta^2}$$

$$\hat\gamma^2 = \frac{\gamma_{l+1}\Delta_l}{\hat\Delta^2 + \dfrac{\gamma_{l+1}\Delta_{l+1}}{\hat\Delta^2 + \Delta_{l+1}}\,\hat\Delta^1 + \hat\Delta^2}$$

$$\hat\delta^1 = \frac{\delta_l \Delta_l}{\hat\Delta^1 + \dfrac{\frac{\delta_l}{\epsilon_l\mu_1}\Delta_{l-1} + \hat\Delta^1}{\Delta_{l-1} + \hat\gamma^1\hat\Delta^1}(\gamma_l\Delta_l - \hat\gamma^1\hat\Delta^1)}$$

$$\hat\delta^2 = \frac{\delta_{l+1}\mu_2\Delta_l}{\hat\Delta^2 + \dfrac{\delta_{l+1}\mu_2\Delta_{l+1}}{\hat\Delta^2 + \hat\gamma^2\Delta_{l+1}}\,\hat\Delta^1 + \hat\Delta^2}$$

$$\hat\epsilon^1 = \frac{\epsilon_l\mu_1\Delta_l}{\hat\Delta^1 + \dfrac{\epsilon_l\mu_1\Delta_{l-1} + \hat\Delta^1}{\hat\gamma^1\Delta_{l-1} + \hat\Delta^1}\,\hat\Delta^2}$$

$$\hat\epsilon^2 = \frac{\epsilon_{l+1}\Delta_l}{\dfrac{\hat\Delta^2 + \frac{\epsilon_{l+1}}{\delta_{l+1}\mu_2}\Delta_{l+1}}{\hat\gamma^2\hat\Delta^2 + \Delta_{l+1}}(\gamma_{l+1}\Delta_l - \hat\gamma^2\hat\Delta^2) + \hat\Delta^2}$$

where the following abbreviations are used

$$\hat\Delta^1 = \hat u - u_l\,, \quad \hat\Delta^2 = u_{l+1} - \hat u\,, \quad \hat u \in\, ]u_l, u_{l+1}[$$

and

$$\mu_1 = \frac{\Delta_l}{\hat\Delta^1 + \dfrac{\epsilon_l\Delta_{l-1} + \hat\Delta^1}{\gamma_l\Delta_{l-1} + \hat\Delta^1}\,\hat\Delta^2}$$

$$\mu_2 = \frac{\Delta_l}{\dfrac{\hat\Delta^2 + \delta_{l+1}\Delta_{l+1}}{\hat\Delta^2 + \gamma_{l+1}\Delta_{l+1}}\,\hat\Delta^1 + \hat\Delta^2}\,.$$

The new design parameters in the quintic case are outlined in Appendix C.

Please note that in (6.5)–(6.9) and (6.11) $\hat u \neq u_l$ is assumed. Only for $n = 3$ the equations (6.5)–(6.9) and (6.11) are valid also for $\hat u = u_l$. But obviously, in the cases $n \geq 4$ the parameter value $\hat u = u_l$ needs a special treatment because at this value the curve is not $C^{n-2}$-continuous so that the construction of the de Casteljau algorithm cannot be embedded. So, for $\hat u = u_l$ (6.6) and (6.7) have to be replaced by

$$\hat\gamma_l = 1\,, \quad \hat\delta_l = \tfrac{1}{\gamma_l}\,, \quad \hat\epsilon_l = 1\,, \quad \hat\rho_l = \tfrac{1}{\gamma_l\delta_l}\,, \quad \hat\sigma_l = \tfrac{1}{\gamma_l}\,, \quad \hat\tau_l = 1$$

$$\hat\gamma_{l+1} = 1\,, \hat\delta_{l+1} = 1\,, \hat\epsilon_{l+1} = \tfrac{1}{\gamma_l}\,, \hat\rho_{l+1} = 1\,, \hat\sigma_{l+1} = \tfrac{1}{\gamma_l}\,, \hat\tau_{l+1} = \tfrac{1}{\gamma_l\delta_l}$$

where $\delta_l = \epsilon_l$ is assumed in the case $n = 5$.

Additionally, for $\hat{u} \in [u_l, u_{l+1}[$ the factors $\hat{\alpha}_{j,n}$ can be obtained to

$$\hat{\alpha}_{l-2,3}(\hat{u}) = F_{l-1}(\gamma_{l-1}\triangle_{l-2} + \triangle_{l-1} + \hat{\gamma}^1\hat{\triangle}^1)$$

$$\hat{\alpha}_{l-1,3}(\hat{u}) = F_l(\gamma_l\triangle_{l-1} + \hat{\triangle}^1)$$

$$\hat{\alpha}_{l,3}(\hat{u}) = F_{l+1}(\gamma_{l+1}\triangle_l - \hat{\gamma}^2\hat{\triangle}^2)$$

$$\hat{\alpha}_{l-3,4}(\hat{u}) = C_{l-1}(\Gamma_{l-1}\triangle_{l-3} + \delta_{l-1}\triangle_{l-2} + \epsilon_{l-1}\triangle_{l-1} + \hat{\hat{\Gamma}}_{l-1}\hat{\triangle}^1)$$

$$\hat{\alpha}_{l-2,4}(\hat{u}) = C_l(\Gamma_l\triangle_{l-2} + \delta_l\triangle_{l-1} + \epsilon_l\mu_1\hat{\triangle}^1)$$

$$\hat{\alpha}_{l-1,4}(\hat{u}) = C_{l+1}(\Gamma_{l+1}\triangle_{l-1} + \delta_{l+1}(\triangle_l - \mu_2\hat{\triangle}^2))$$

$$\hat{\alpha}_{l,4}(\hat{u}) = C_{l+2}(\Gamma_{l+2}\triangle_l - \hat{\Gamma}_{l+3}\hat{\triangle}^2)$$

$$\hat{\alpha}_{l-4,5}(\hat{u}) = A_{l-2}(\Omega_{l-2}\triangle_{l-4} + \Psi_{l-2}\triangle_{l-3} + \triangle_{l-2} + \overline{\Psi}_{l-2}\triangle_{l-1} + \hat{\overline{\Omega}}_{l-2}\hat{\triangle}^1)$$

$$\hat{\alpha}_{l-3,5}(\hat{u}) = A_{l-1}(\Omega_{l-1}\triangle_{l-3} + \Psi_{l-1}\triangle_{l-2} + \triangle_{l-1} + \overline{\Psi}_{l-1}\mu_{15}\hat{\triangle}^1)$$

$$\hat{\alpha}_{l-2,5}(\hat{u}) = A_l(\Omega_l\triangle_{l-2} + \Psi_l\triangle_{l-1} + \mu_{16}\hat{\triangle}^1)$$

$$\hat{\alpha}_{l-1,5}(\hat{u}) = A_{l+1}(\Omega_{l+1}\triangle_{l-1} + \Psi_{l+1}(\triangle_l - \mu_{17}\hat{\triangle}^2))$$

$$\hat{\alpha}_{l,5}(\hat{u}) = A_{l+2}(\Omega_{l+2}\triangle_l - \hat{\Omega}_{l+3}\hat{\triangle}^2)$$

where some abbreviations are used which have been introduced in Appendix A, B, and C. One example for the cases $n = 4$ (respectively, $n = 5$) can be seen in Figs. 6.7 and 6.8.

*Remarks.* (1) Algebraically a solution of the inserting problem can be formulated as follows. The formula (6.10) is a direct result of the identities

$$
\begin{aligned}
G^n_{l-n}(u) &= & \hat{G}^n_{l-n}(u) &+ (1 - \hat{\alpha}_{l-n+1,n}(\hat{u})) & \hat{G}^n_{l-n+1}(u) \\
G^n_{l-n+1}(u) &= \hat{\alpha}_{l-n+1,n}(\hat{u})\, \hat{G}^n_{l-n+1}(u) &+ (1 - \hat{\alpha}_{l-n+2,n}(\hat{u})) & \hat{G}^n_{l-n+2}(u) \\
&\vdots & \vdots & \\
G^n_{l-1}(u) &= \hat{\alpha}_{l-1,n}(\hat{u})\, \hat{G}^n_{l-1}(u) &+ & (1 - \hat{\alpha}_{l,n}(\hat{u}))\, \hat{G}^n_l(u) \\
G^n_l(u) &= \hat{\alpha}_{l,n}(\hat{u})\, \hat{G}^n_l(u) &+ & \hat{G}^n_{l+1}(u)
\end{aligned}
$$

(6.12)

for $u \in [u_{l-n}, u_{l+n+1}]$. Now using the properties of partition of unity of the $G^n_i$ and $\hat{G}^n_i$, we can solve the linear system (6.12) at $u = \hat{u}$. The solution is

$$(6.13) \quad \hat{\alpha}_{r,n}(\hat{u}) = \frac{\sum_{j=r}^l G^n_j(\hat{u}) - \sum_{j=r+1}^l \hat{G}^n_j(\hat{u})}{\hat{G}^n_r(\hat{u})} \quad (r = l - n + 1, \cdots, l)\,.$$

We get an explicit expression of (6.13) by inserting the Bézier ordinates of $G^n_j$ and $\hat{G}^n_j$, but obviously (6.13) is still depending on the unknowns of (6.6) and (6.8).

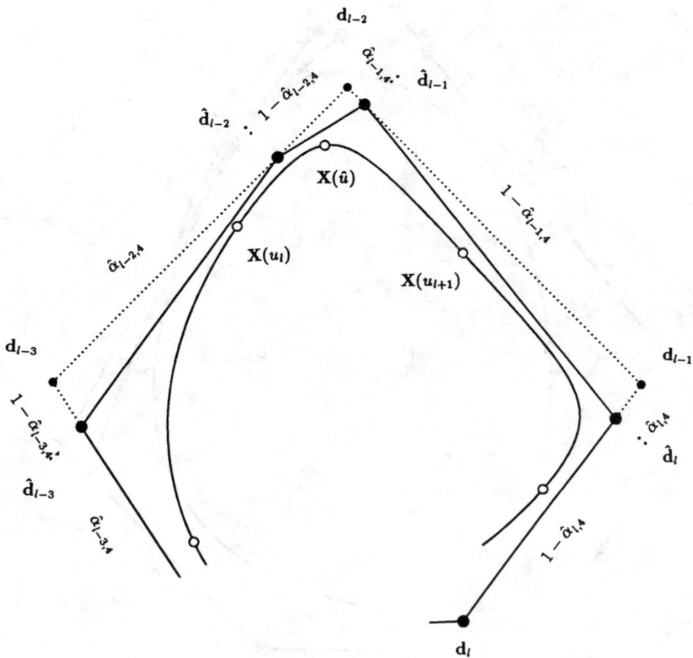

FIG. 6.7.  *Insertion of a new knot in the quartic case.*

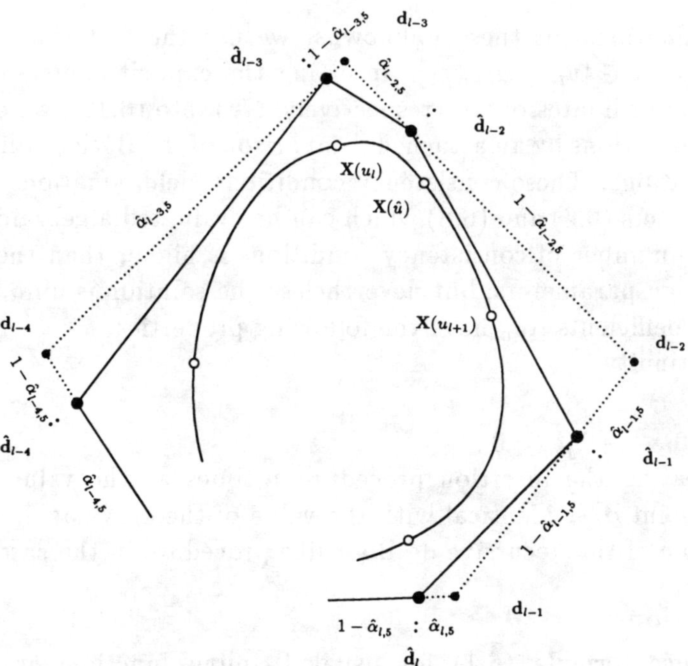

FIG. 6.8.  *Insertion of a new knot in the quintic case.*

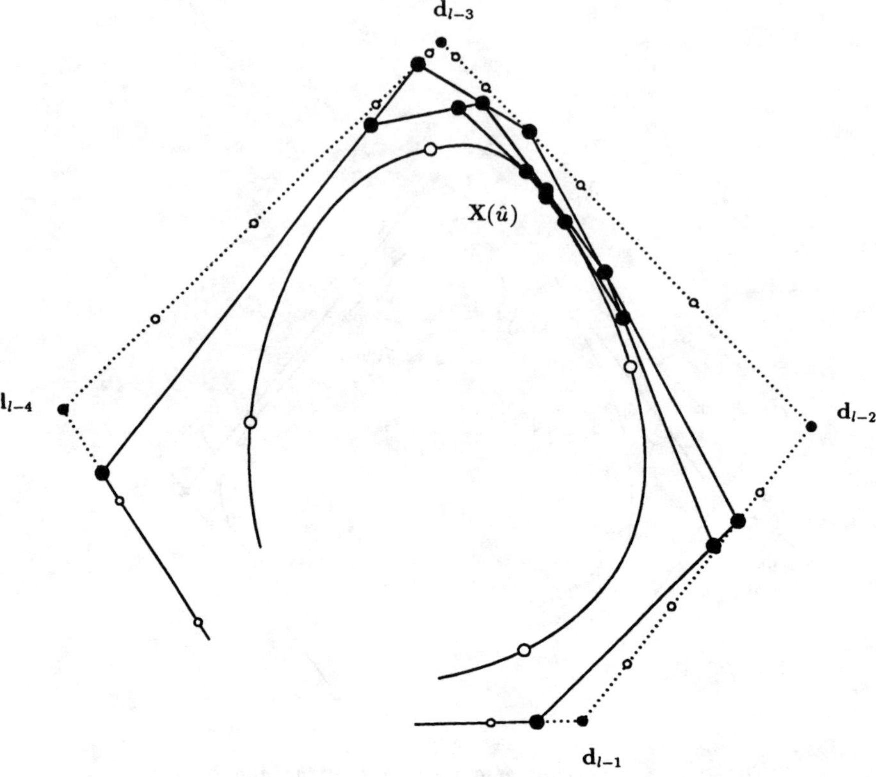

FIG. 6.9. *Insertion of $\hat{u}$ n-times (de Boor-like algorithm), $n = 5$.*

To get equations for these unknowns, we use the fact that (6.12) must be valid for all $u \in [u_{l-n}, u_{l+n+1}]$. Inserting the explicit expression of (6.13) and the Bézier ordinates of $G_i^n$ (respectively, $\hat{G}_i^n$) into (6.12), we obtain some consistency conditions by an attached comparison of the Bézier ordinates in the resulting equations. These consistency conditions yield equations for the new design parameters (6.6) and (6.8) which can be evaluated algebraically. Please note that the number of consistency conditions is higher than the number of unknown design parameters, but nevertheless the solution is unique.

(2) The coefficients $\hat{\alpha}_{j,n}$ have the following properties
- $C^1$-continuity

- $\hat{\alpha}_{j,n} \geq 0$ , $1 - \hat{\alpha}_{j,n} \geq 0$ .

(3) Repeating the insertion procedure $n$-times at the value $\hat{u}$, the last determined point $\mathbf{d}_l$ is identical with the value of the curve at $\hat{u}$. See Fig. 6.9 for an example of this recursive de Boor-like procedure in the case $n = 5$.

## 6.4. Recursion

The recurrence formula (6.1) for usual B-spline functions is the second algorithm we will adapt to geometric continuous B-splines. First we realize that the insertion coefficients $\hat{\alpha}_{i,n}$ derived in the previous section cannot be used for such a recursion.

On the other hand, in 1985 Goodman and Unsworth [12] published a Mansfield-like formula

$$G_i^3(u) = \alpha_i^3 \cdot N_i^2(u) + \beta_{i+1}^3 \cdot N_{i+1}^2(u)$$

which uses linear coefficients $\alpha_i^3$ and $\beta_i^3$. But this recursion is not desirable because the coefficients build no convex combination $(\alpha_i^3 + \beta_i^3 \neq 1)$.

Therefore to get a recursion for the geometric B-spline function $G_i^n$ based on a convex combination with nonlinear coefficients, we use the following segmentwise formulation for $u \in [u_l, u_{l+1}]$

$$G_{i,l-i}^n(u) = \alpha_{i,n}(u) \cdot G_{i,l-i}^{n-1}(u) + (1 - \alpha_{i+1,n}(u)) \cdot G_{i+1,l-i-1}^{n-1}(u) \qquad (i = l-n, \cdots, l)$$

where the local support basis functions $G_i^n(u)$ are related to the $G_i^{n-1}(u)$ using some unknown coefficients $\alpha_{i,n}(u)$. Interpreting these equations again as a linear system, we get a solution of the $\alpha_{i,n}(u)$ similar to the one in the last section if we use the properties by which the $G_i^n$ and $G_i^{n-1}$ build partitions of unity:

$$(6.14) \quad \alpha_{r,n}(u) = \frac{\sum_{j=r}^{l} G_{j,l-j}^n(u) - \sum_{j=r+1}^{l} G_{j,l-j}^{n-1}(u)}{G_{r,l-r}^{n-1}(u)} \qquad (r = l-n, \cdots, l) .$$

Now, inserting (6.4), i.e., the representations of the $G_i^n$ and $G_i^{n-1}$, in Bézier ordinates into the quotient (6.14) and subsequently executing a degree elevation of the Bézier ordinates of the $G_i^{n-1}$ in the numerator, we get for $\alpha_{r,n}(u)$, $(r = l-n, \cdots, l)$

$$\frac{\sum_{k=0}^{n} \left( b_{nl+k,r} + \sum_{j=r+1}^{l} \left( b_{nl+k,j} - \frac{k}{n} b_{(n-1)l+k-1,j} - \frac{n-k}{n} b_{(n-1)l+k,j} \right) \right) B_k^n \left( \frac{u-u_l}{\Delta_l} \right)}{\sum_{k=0}^{n-1} b_{(n-1)l+k,r} B_k^{n-1} \left( \frac{u-u_l}{\Delta_l} \right)} .$$

Explicit expressions of the $\alpha_{r,n}(u)$ can be determined by inserting the factors $b_{nl+s,j}$ of degree $n$ and $b_{(n-1)l+s,j}$ of degree $n-1$ which are given in Appendix B.

For $n = 1$ and $n = 2$, we obtain the well known coefficients of the usual $C^1$-continuous B-splines:

$$\begin{aligned} \alpha_{r,1}(u) &= \frac{u-u_r}{\Delta_r} , & u &\in [u_r, u_{r+1}] , \\ \alpha_{r,2}(u) &= \frac{u-u_r}{\Delta_r + \Delta_{r+1}} , & u &\in [u_r, u_{r+2}] . \end{aligned}$$

Naturally, the coefficients $\alpha_{i,n}$ are more complicated in the cases $n = 3$, $n = 4$, and $n = 5$, because of the many apparent design parameters. For instance in the cubic case, we obtain

$$\alpha_{r,3}(u) = \begin{cases} \dfrac{\gamma_{r+1}(u-u_r)}{\gamma_{r+1}\triangle_r + \triangle_{r+1} + \gamma_{r+2}\triangle_{r+2}}, & u \in [u_r, u_{r+1}[ \\[2em] \dfrac{\sum_{k=0}^{3} g_k B_k^3\left(\frac{u-u_{r+1}}{\triangle_{r+1}}\right)}{\sum_{k=0}^{2} h_k B_k^2\left(\frac{u-u_{r+1}}{\triangle_{r+1}}\right)}, & u \in [u_{r+1}, u_{r+2}[ \\[2em] \dfrac{\gamma_{r+1}\triangle_r + \triangle_{r+1} + \gamma_{r+2}(u-u_{r+2})}{\gamma_{r+1}\triangle_r + \triangle_{r+1} + \gamma_{r+2}\triangle_{r+2}}, & u \in [u_{r+2}, u_{r+3}] \end{cases}$$

where

$$g_0 = \frac{\triangle_r}{\triangle_r + \triangle_{r+1}} \cdot \frac{\gamma_{r+1}\triangle_r}{\gamma_{r+1}\triangle_r + \triangle_{r+1} + \gamma_{r+2}\triangle_{r+2}}$$

$$g_1 = \frac{\gamma_{r+1}\triangle_r}{\gamma_{r+1}\triangle_r + \triangle_{r+1} + \gamma_{r+2}\triangle_{r+2}}$$

$$g_2 = \frac{\gamma_{r+1}\triangle_r + \triangle_{r+1}}{\gamma_{r+1}\triangle_r + \triangle_{r+1} + \gamma_{r+2}\triangle_{r+2}} - \frac{1}{3} \cdot \frac{\triangle_{r+1}}{\triangle_{r+1} + \triangle_{r+2}}$$

$$g_3 = \frac{\triangle_{r+2}}{\triangle_{r+1} + \triangle_{r+2}} \cdot \frac{\gamma_{r+1}\triangle_r + \triangle_{r+1}}{\gamma_{r+1}\triangle_r + \triangle_{r+1} + \gamma_{r+2}\triangle_{r+2}}$$

$$h_0 = \frac{\triangle_r}{\triangle_r + \triangle_{r+1}}$$

$$h_1 = 1$$

$$h_2 = \frac{\triangle_{r+2}}{\triangle_{r+1} + \triangle_{r+2}}.$$

In Figs. 6.10–6.12 examples of the coefficients in the cases $n = 3$ to $n = 5$ can be seen, where the dotted lines show the usual coefficients from (6.1).

*Remarks.* (1) The first and last part of $\alpha_{r,n}(u)$ are linear polynomials. The middle parts are generally quotients of polynomials of degree $n$ over polynomials of degree $n - 1$.

(2) $\alpha_{r,n}(u)$ is $C^1$-continuous.

(3) $\alpha_{r,n}(u)$ can be negative. For instance, in the case $n = 3$ because of the apparent difference in the second Bézier value $g_2$ of the numerator.

## 6.5.  De Boor-Like Algorithm

Using the yielded recurrence formula of geometric continuous B-splines, we are able to formulate an de Boor-like algorithm, which is deduced in the same way as the usual de Boor algorithm from the usual recurrence formula:

$$\mathbf{d}_r^j = \alpha_{r,n-j+1}(u) \cdot \mathbf{d}_r^{j-1} + (1 - \alpha_{r,n-j+1}(u)) \cdot \mathbf{d}_{r-1}^{j-1} \; ; \quad \mathbf{d}_r^0 = \mathbf{d}_r$$

$$u \in [u_l, u_{l+1}[ \; ; \quad (r = l - n + j, \cdots, l) \; ; \quad (j = 1, \cdots, n).$$

The last determined point $\mathbf{d}_l^n$ of this recursion is identical to the point $\mathbf{X}(u)$ of the curve at $u$. Figs. 6.13–6.15 show examples of this algorithm in the cases $n = 3$, $n = 4$, and $n = 5$.

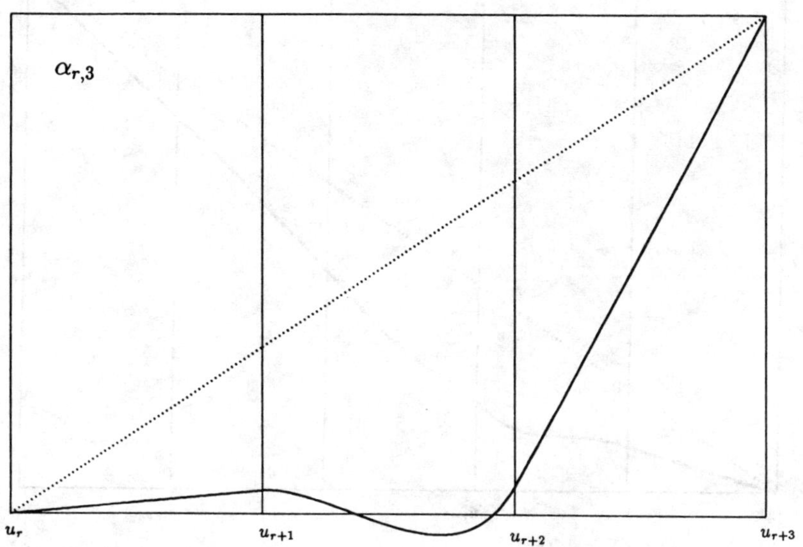

FIG. 6.10. $\alpha_{r,3}(u)$ in the $GC^2$-continuity case.

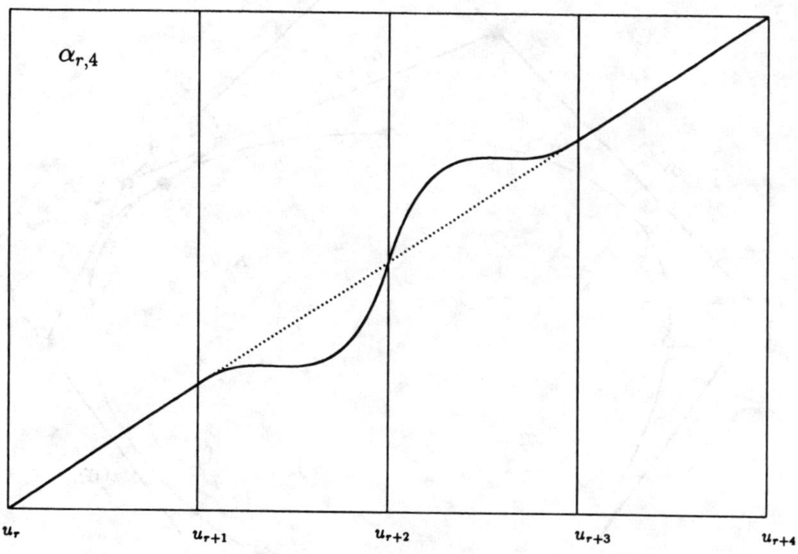

FIG. 6.11. $\alpha_{r,4}(u)$ in the $GC^3$-continuity case.

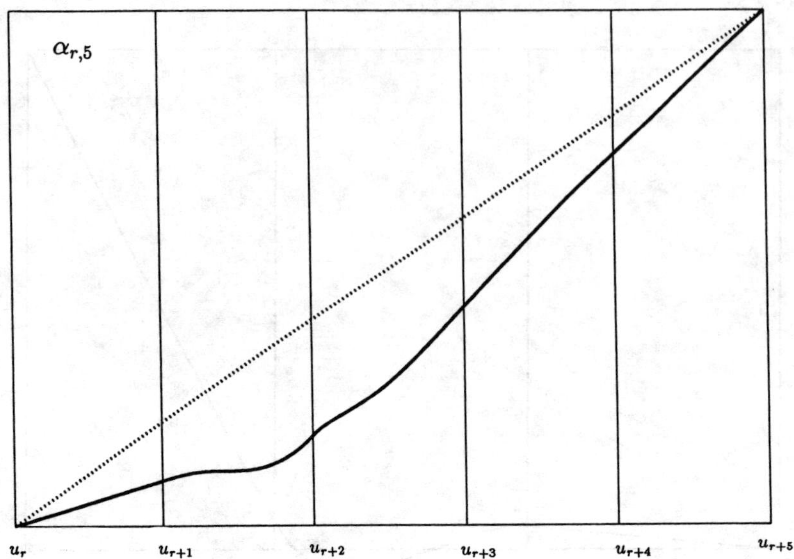

FIG. 6.12.  $\alpha_{r,5}(u)$ *in the* $GC'^4$-*continuity case.*

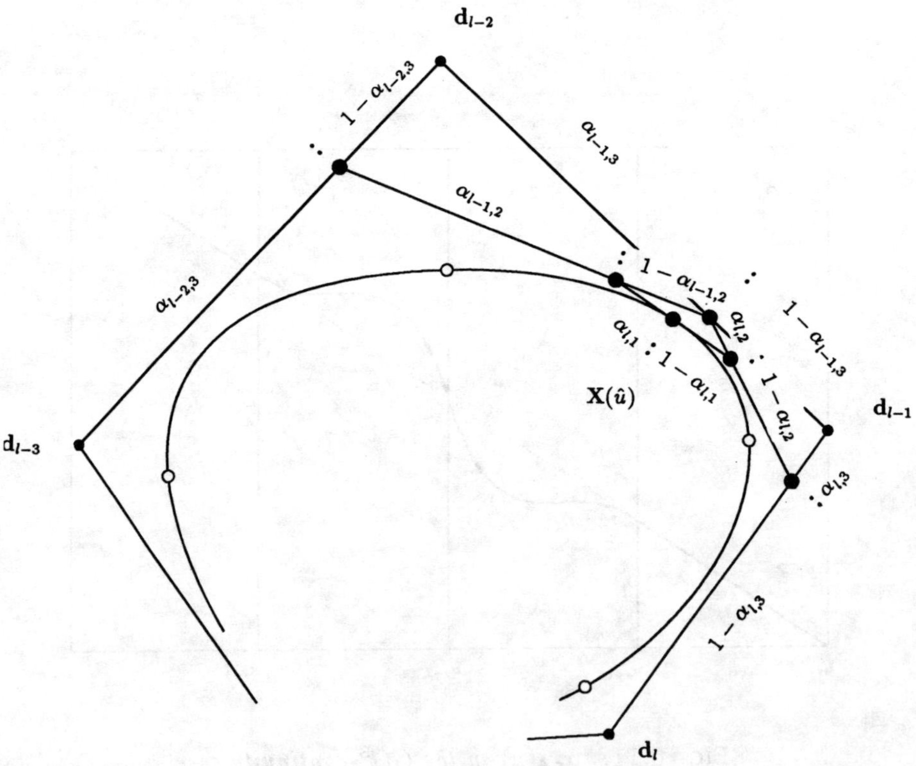

FIG. 6.13.  *De Boor-like algorithm* $(n = 3)$.

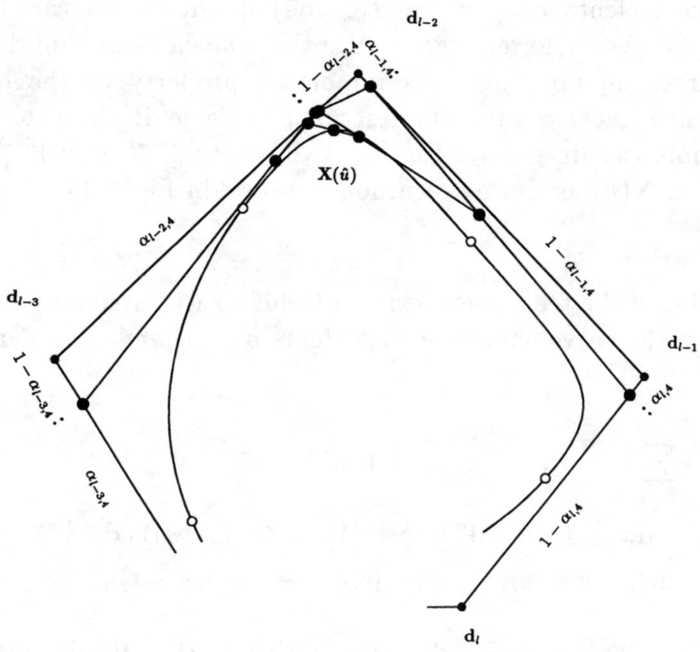

FIG. 6.14.  *De Boor-like algorithm* $(n = 4)$.

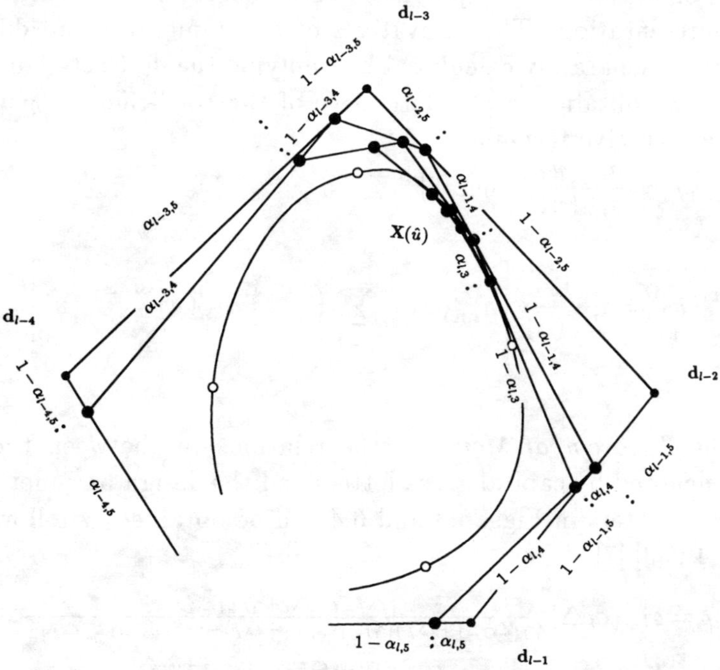

FIG. 6.15.  *De Boor-like algorithm* $(n = 5)$.

The main advantage of this new recursive algorithm is that all the appearing coefficients $\alpha_{i,n}$ in the (maybe) quinitic case can be used for algorithms of curves of lower degree. All other known algorithms for geometric B-spline curves do not have this important property as the de Boor-like algorithm using linear coefficients described by Seidel in the cubic case [19].

Please note that in general the last two points $\mathbf{d}_l^{n-1}$ and $\mathbf{d}_{l-1}^{n-1}$ are not on the tangent in $\mathbf{X}(u)$ as can be seen, for instance, in Fig. 6.14.

### 6.6. Derivatives

Differentiation of the recurrence formula yields to a recursive algorithm for the derivatives of the curve using the coefficients $\alpha_{i,n}(u)$ and their derivatives :

$$
\begin{aligned}
\mathbf{d}_r^{m,j} &= \sum_{s=1}^{m} \binom{m}{s} \cdot \alpha_{r,n-j+1}^{(s)}(u) \cdot [\mathbf{d}_r^{m-s,j-1} - \mathbf{d}_{r-1}^{m-s,j-1}] \\
&\quad + \alpha_{r,n-j+1}(u) \cdot \mathbf{d}_r^{m,j-1} + (1 - \alpha_{r,n-j+1}(u)) \cdot \mathbf{d}_{r-1}^{m,j-1} \\
\mathbf{d}_r^{0,0} &= \mathbf{d}_r ; \quad u \in [u_l, u_{l+1}] ; \quad (r = l - n + j, \cdots, l) ; \quad (j = 1, \cdots, n) .
\end{aligned}
$$

Again the calculated point $\mathbf{d}_l^{m,n}$ is identical to the $m$th derivative $\mathbf{X}^{(m)}(u)$ of the curve at $u$.

The problem of calculating the derivatives of the coefficients $\alpha_{i,n}$ is also solvable. As already said, these coefficients are in general quotients of polynomials of degree $n$ over polynomials of degree $n - 1$ which are given in Bézier representation. The derivatives of the numerator and denominator can therefore be separately calculated by applying the de Casteljau algorithm. Subsequently we obtain the $r$th derivative of the coefficient $\alpha_{i,n}(u)$ by using the following recursive formula.

Let $\alpha_{i,n}(u) = \dfrac{n_{i,n}(u)}{d_{i,n}(u)}$, then [9]

$$
\alpha_{i,n}^{(r)}(u) = \frac{1}{d_{i,n}(u)} \left[ n_{i,n}^{(r)}(u) - \sum_{j=1}^{r} \binom{r}{j} d_{i,n}^{(j)}(u) n_{i,n}^{(r-j)}(u) \right] .
$$

### Appendix A

Applying the *Theorem of Menelao*, the relationships between the auxiliary parameters denoted by capital greek letters and the design parameters denoted by small greek letters in Figs. 6.3 and 6.4 can be obtained as follows

(a) $n = 4 :$ [3],[7]

$$
\overline{\Gamma}_I = \delta_{I+1}\epsilon_I\Delta_I \frac{\epsilon_I\Delta_{I-1}+\Delta_I+\gamma_{I+1}\Delta_{I+1}}{(\gamma_I\Delta_{I-1}+\Delta_I)(\gamma_{I+1}\Delta_{I+1}+\Delta_I)-\epsilon_I\delta_{I+1}\Delta_{I-1}\Delta_{I+1}}
$$

$$
\Gamma_I = \delta_I\epsilon_{I-1}\Delta_{I-1} \frac{\gamma_{I-1}\Delta_{I-2}+\Delta_{I-1}+\delta_I\Delta_I}{(\gamma_{I-1}\Delta_{I-2}+\Delta_{I-1})(\gamma_I\Delta_I+\Delta_{I-1})-\epsilon_{I-1}\delta_I\Delta_{I-2}\Delta_I}
$$

(b) $n = 5 :$ [7],[8]

$$\Theta_I = \tau_{I-1}\delta_I \Delta_{I-1} \frac{\epsilon_{I-1}\Delta_{I-2}+\Delta_{I-1}+\delta_I\Delta_I}{(\epsilon_{I-1}\Delta_{I-2}+\Delta_{I-1})(\Delta_{I-1}+\gamma_I\Delta_I)-\tau_{I-1}\delta_I\Delta_{I-2}\Delta_I}$$

$$\overline{\Theta}_I = \epsilon_I\rho_{I+1}\Delta_I \frac{\epsilon_I\Delta_{I-1}+\Delta_I+\delta_{I+1}\Delta_{I+1}}{(\gamma_I\Delta_{I-1}+\Delta_I)(\Delta_I+\delta_{I+1}\Delta_{I+1})-\rho_{I+1}\epsilon_I\Delta_{I-1}\Delta_{I+1}}$$

$$\overline{\Lambda}_I = \tau_I\delta_{I+1}\Delta_I \frac{\tau_I\Delta_{I-1}+\Delta_I+\gamma_{I+1}\Delta_{I+1}}{(\epsilon_I\Delta_{I-1}+\Delta_I)(\Delta_I+\gamma_{I+1}\Delta_{I+1})-\tau_I\delta_{I+1}\Delta_{I-1}\Delta_{I+1}}$$

$$\Phi_I = \epsilon_{I-1}\rho_I\Delta_{I-1} \frac{\gamma_{I-1}\Delta_{I-2}+\Delta_{I-1}+\rho_I\Delta_I}{(\gamma_{I-1}\Delta_{I-2}+\Delta_{I-1})(\Delta_{I-1}+\delta_I\Delta_I)-\epsilon_{I-1}\rho_I\Delta_{I-2}\Delta_I}$$

$$\Lambda_I = \sigma_I\Theta_I\Delta_{I-1} \frac{\Phi_I\Delta_{I-2}+\rho_I\Delta_{I-1}+\sigma_I\Delta_I}{(\Phi_I\Delta_{I-2}+\rho_I\Delta_{I-1})(\delta_I\Delta_{I-1}+\epsilon_I\Delta_I)-\sigma_I\Theta_I\Delta_{I-2}\Delta_I}$$

$$\overline{\Phi}_I = \sigma_I\overline{\Theta}_I\Delta_I \frac{\sigma_I\Delta_{I-1}+\tau_I\Delta_I+\overline{\Lambda}_I\Delta_{I+1}}{(\delta_I\Delta_{I-1}+\epsilon_I\Delta_I)(\tau_I\Delta_I+\overline{\Lambda}_I\Delta_{I+1})-\sigma_I\overline{\Theta}_I\Delta_{I-1}\Delta_{I+1}}$$

$$\Psi_I = \sigma_I\Delta_I \frac{\delta_I\Delta_{I-1}+\epsilon_I\Delta_I+\overline{\Theta}_I\Delta_{I+1}}{(\delta_I\Delta_{I-1}+\epsilon_I\Delta_I)(\tau_I\Delta_I+\overline{\Lambda}_I\Delta_{I+1})-\sigma_I\overline{\Theta}_I\Delta_{I-1}\Delta_{I+1}}$$

$$\overline{\Psi}_I = \sigma_{I+1}\Delta_I \frac{\Theta_{I+1}\Delta_{I-1}+\delta_{I+1}\Delta_I+\epsilon_{I+1}\Delta_{I+1}}{(\Phi_{I+1}\Delta_{I-1}+\rho_{I+1}\Delta_I)(\delta_{I+1}\Delta_I+\epsilon_{I+1}\Delta_{I+1})-\sigma_{I+1}\Theta_{I+1}\Delta_{I-1}\Delta_{I+1}}$$

$$\overline{\Omega}_{I-1} = \frac{\overline{\Phi}_I(\Lambda_I\Delta_{I-2}+\sigma_I\Delta_{I-1}+\tau_I\Delta_I+\overline{\Lambda}_I\Delta_{I+1})(\Delta_{I-1}+\overline{\Psi}_{I-1}\Delta_I)}{(\Phi_I\Delta_{I-2}+\rho_I\Delta_{I-1}+\sigma_I\Delta_I)(\sigma_I\Delta_{I-1}+\tau_I\Delta_I+\overline{\Lambda}_I\Delta_{I+1})-\overline{\Phi}_I\Lambda_I\Delta_{I-2}\Delta_{I+1}}$$

$$\Omega_I = \frac{\Lambda_I(\Phi_I\Delta_{I-2}+\rho_I\Delta_{I-1}+\sigma_I\Delta_I+\overline{\Phi}_I\Delta_{I+1})(\Psi_I\Delta_{I-1}+\Delta_I)}{(\Phi_I\Delta_{I-2}+\rho_I\Delta_{I-1}+\sigma_I\Delta_I)(\sigma_I\Delta_{I-1}+\tau_I\Delta_I+\overline{\Lambda}_I\Delta_{I+1})-\overline{\Phi}_I\Lambda_I\Delta_{I-2}\Delta_{I+1}}$$

## Appendix B

In this Appendix the remaining factors $b_{nl+i,l-j}$ from (6.3) are explicitly given
in the cases $n = 3$ to $n = 5$. The following equations can easily be derived
from Figs. 6.2 to 6.4:

$$\begin{aligned}
b_{3l+1,l-1} &= F_l\gamma_l\Delta_{l-1} \\
b_{3l+2,l-1} &= F_l(\gamma_l\Delta_{l-1} + \Delta_l) \\
b_{3l,l-1} &= I_l\Delta_{l-1}b_{3l+1,l-1} \\
b_{3l,l-3} &= I_l\Delta_l b_{3(l-1)+2,l-3}
\end{aligned}$$

$$\begin{aligned}
b_{4l+2,l-1} &= E_l C_{l+1}\Gamma_{l+1}\Delta_{l-1}(\gamma_l\Delta_{l-1}+\Delta_l) \\
b_{4l+2,l-3} &= E_l C_l\delta_{l+1}\Delta_{l+1}(\epsilon_l\Delta_l+\overline{\Gamma}_l\Delta_{l+1}) \\
b_{4l+1,l-1} &= G_l\gamma_l\Delta_{l-1}b_{4l+2,l-1} \\
b_{4l+1,l-3} &= G_l(\gamma_l\Delta_{l-1}b_{4l+2,l-3}+C_l\Delta_l(\epsilon_l+\overline{\Gamma}_l\Delta_{l+1})) \\
b_{4l+3,l-1} &= H_{l+1}(C_{l+1}\Delta_l(\Gamma_{l+1}\Delta_{l-1}+\delta_{l+1}\Delta_l)+\gamma_{l+1}\Delta_{l+1}b_{4l+2,l-1}) \\
b_{4l+3,l-3} &= H_{l+1}\gamma_{l+1}\Delta_{l+1}b_{4l+2,l-3} \\
b_{4l,l-1} &= I_l\Delta_{l-1}b_{4l+1,l-1} \\
b_{4l,l-3} &= I_l(\Delta_{l-1}b_{4l+1,l-3}+\Delta_l b_{4(l-1)+3,l-3}) \\
b_{4l,l-4} &= I_l\Delta_l b_{4(l-1)+3,l-4}
\end{aligned}$$

$$b_{5l+2,l-1} = A_{l+1}B_{l+1}D_l\Theta_{l+1}\epsilon_l\triangle_{l-1}^2(\Omega_{l+1}\triangle_{l-1}+\Psi_{l+1}\triangle_l)$$

$$b_{5l+2,l-3} = D_l(A_lB_{l+1}\epsilon_l\triangle_{l-1}(\delta_{l+1}\triangle_l+\epsilon_{l+1}\triangle_{l+1}\overline{\Theta}_{l+1}\triangle_{l+2})(\overline{\Psi}_l\triangle_{l+1}+\overline{\Omega}_l\triangle_{l+2})$$
$$+B_l(\triangle_l+\delta_{l+1}\triangle_{l+1})(A_l(\Theta_l\triangle_{l-2}+\delta_l\triangle_{l-1}+\epsilon_l\triangle_l)$$
$$(\triangle_l+\overline{\Psi}_l\triangle_{l+1}+\overline{\Omega}_l\triangle_{l+2})$$
$$+A_{l-1}\overline{\Theta}_l\triangle_{l+1}(\Omega_{l-1}\triangle_{l-3}+\Psi_{l-1}\triangle_{l-2}+\triangle_{l-1})))$$

$$b_{5l+2,l-4} = A_{l-1}B_lD_l\overline{\Theta}_l\delta_{l+1}\triangle_{l+1}^2(\overline{\Psi}_{l-1}\triangle_l+\overline{\Omega}_{l-1}\triangle_{l+1})$$

$$b_{5l+3,l-1} = A_{l+1}B_{l+1}D_l\Theta_{l+1}\triangle_{l-1}(\epsilon_l\triangle_{l-1}+\triangle_l)(\Omega_{l+1}\triangle_{l-1}+\Psi_{l+1}\triangle_l)$$

$$b_{5l+3,l-3} = D_l(A_lB_{l+1}(\epsilon_l\triangle_{l-1}+\triangle_l)(\delta_{l+1}\triangle_l+\epsilon_{l+1}\triangle_{l+1}\overline{\Theta}_{l+1}\triangle_{l+2})$$
$$(\overline{\Psi}_l\triangle_{l+1}+\overline{\Omega}_l\triangle_{l+2})$$
$$+B_l\delta_{l+1}\triangle_{l+1}(A_l(\Theta_l\triangle_{l-2}+\delta_l\triangle_{l-1}+\epsilon_l\triangle_l)(\triangle_l+\overline{\Psi}_l\triangle_{l+1}+\overline{\Omega}_l\triangle_{l+2})$$
$$+A_{l-1}\overline{\Theta}_l\triangle_{l+1}(\Omega_{l-1}\triangle_{l-3}+\Psi_{l-1}\triangle_{l-2}+\triangle_{l-1})))$$

$$b_{5l+3,l-4} = A_{l-1}B_lD_l\overline{\Theta}_l\delta_{l+1}\triangle_{l+1}^2(\overline{\Psi}_{l-1}\triangle_l+\overline{\Omega}_{l-1}\triangle_{l+1})$$

$$b_{5l+1,l-1} = G_l\gamma_l\triangle_{l-1}b_{5l+2,l-1}$$

$$b_{5l+1,l-3} = G_l(\gamma_l\triangle_{l-1}b_{5l+2,l-3}+A_lB_l\triangle_l(\Theta_l\triangle_{l-2}+\delta_l\triangle_{l-1})$$
$$(\triangle_l+\overline{\Psi}_l\triangle_{l+1}+\overline{\Omega}_l\triangle_{l+2})$$
$$+A_{l-1}B_l\triangle_l(\epsilon_l\triangle_l+\overline{\Theta}_l\triangle_{l+1})(\Omega_{l-1}\triangle_{l-3}+\Psi_{l-1}\triangle_{l-2}+\triangle_{l-1}))$$

$$b_{5l+1,l-4} = G_l(\gamma_l\triangle_{l-1}b_{5l+2,l-4}+A_{l-1}B_l\triangle_l(\epsilon_l\triangle_l+\overline{\Theta}_l\triangle_{l+1})$$
$$(\overline{\Psi}_{l-1}\triangle_l+\overline{\Omega}_{l-1}\triangle_{l+1}))$$

$$b_{5l+4,l-1} = H_{l+1}(\gamma_{l+1}\triangle_{l+1}b_{5l+3,l-1}+A_{l+1}B_{l+1}\triangle_l(\Theta_{l+1}\triangle_{l-1}+\delta_{l+1}\triangle_l)\cdot$$
$$(\Omega_{l+1}\triangle_{l-1}+\Psi_{l+1}\triangle_l))$$

$$b_{5l+4,l-3} = H_{l+1}(\gamma_{l+1}\triangle_{l+1}b_{5l+3,l-3}+A_lB_{l+1}\triangle_l(\epsilon_{l+1}\triangle_{l+1}+\overline{\Theta}_{l+1}\triangle_{l+2})\cdot$$
$$(\overline{\Psi}_l\triangle_{l+1}+\overline{\Omega}_l\triangle_{l+2}))$$

$$b_{5l+4,l-4} = H_{l+1}\gamma_{l+1}\triangle_{l+1}b_{5l+3,l-4}$$

$$b_{5l,l-1} = I_l\triangle_{l-1}b_{5l+1,l-1}$$

$$b_{5l,l-3} = I_l(\triangle_{l-1}b_{5l+1,l-3}+\triangle_lb_{5(l-1)+4,l-3})$$

$$b_{5l,l-4} = I_l(\triangle_{l-1}b_{5l+1,l-4}+\triangle_lb_{5(l-1)+4,l-4})$$

$$b_{5l,l-5} = I_l\triangle_lb_{5(l-1)+4,l-5}$$

where the following abbreviations are used

$$A_l = (\Omega_l\triangle_{l-2}+\Psi_l\triangle_{l-1}+\triangle_l+\overline{\Psi}_l\triangle_{l+1}+\overline{\Omega}_l\triangle_{l+2})^{-1}$$
$$B_l = (\Theta_l\triangle_{l-2}+\delta_l\triangle_{l-1}+\epsilon_l\triangle_l+\overline{\Theta}_l\triangle_{l+1})^{-1}$$
$$C_l = (\Gamma_l\triangle_{l-2}+\delta_l\triangle_{l-1}+\epsilon_l\triangle_l+\overline{\Gamma}_l\triangle_{l+1})^{-1}$$
$$D_l = (\epsilon_l\triangle_{l-1}+\triangle_l+\delta_{l+1}\triangle_{l+1})^{-1}$$
$$E_l = (\gamma_l\triangle_{l-1}+\triangle_l+\delta_{l+1}\triangle_{l+1})^{-1}$$
$$F_l = (\gamma_l\triangle_{l-1}+\triangle_l+\gamma_{l+1}\triangle_{l+1})^{-1}$$
$$G_l = (\gamma_l\triangle_{l-1}+\triangle_l)^{-1}$$
$$H_l = (\triangle_{l-1}+\gamma_l\triangle_l)^{-1}$$

$$I_l = (\triangle_{l-1} + \triangle_l)^{-1}$$

## Appendix C

In this section of the Appendix we want to give explicit expressions of the new design parameters in the quintic case additionally to (6.11), which result from inserting a new knot $\hat u \in ]u_l, u_{l+1}[$.

$$\hat\rho^1 = \frac{\rho_l \triangle_l}{\hat\triangle^1 + \dfrac{\frac{\rho_l}{\sigma_l\mu_3}\triangle_{l-1} + \hat\triangle^1}{\triangle_{l-1} + \hat\delta^1\hat\triangle^1}(\delta_l\triangle_l - \hat\delta^1\hat\triangle^1)}$$

$$\hat\rho^2 = \frac{\rho_{l+1}\mu_8\triangle_l}{\hat\triangle^2 + \dfrac{\rho_{l+1}\mu_8\triangle_{l+1}}{\hat\triangle^2 + \hat\delta^2\triangle_{l+1}}\hat\triangle^1 + \hat\triangle^2}$$

$$\hat\sigma^1 = \frac{\sigma_l\mu_3\triangle_l}{\hat\triangle^1 + \mu_6\dfrac{\mu_5\triangle_{l-1} + \hat\triangle^1}{\hat\delta^1\triangle_{l-1} + \hat\epsilon^1\hat\triangle^1}\hat\triangle^2}$$

$$\hat\sigma^2 = \frac{\sigma_{l+1}\mu_7\triangle_l}{\mu_9\dfrac{\hat\triangle^2 + \mu_{10}\triangle_{l+1}}{\hat\delta^2\hat\triangle^2 + \hat\epsilon^2\triangle_{l+1}}\hat\triangle^1 + \hat\triangle^2}$$

$$\hat\tau^1 = \frac{\tau_l\mu_4\triangle_l}{\hat\triangle^1 + \dfrac{\tau_l\mu_4\triangle_{l-1} + \hat\triangle^1}{\hat\epsilon^1\triangle_{l-1} + \hat\triangle^1}\hat\triangle^2}$$

$$\hat\tau^2 = \frac{\tau_{l+1}\triangle_l}{\hat\triangle^2 + \dfrac{\frac{\tau_{l+1}}{\sigma_{l+1}\mu_7}\triangle_{l+1}}{\hat\epsilon^2\hat\triangle^2 + \triangle_{l+1}}(\epsilon_{l+1}\triangle_l - \hat\epsilon^2\hat\triangle^2) + \hat\triangle^2}$$

where

$$\mu_3 = \frac{\triangle_l}{\hat\triangle^1 + \dfrac{\epsilon_l}{\tau_l\mu_{11}}\dfrac{\sigma_l\triangle_{l-1} + \tau_l\mu_{11}\hat\triangle^1}{\delta_l\triangle_{l-1} + \epsilon_l\mu_1\hat\triangle^1}(\triangle_l - \mu_1\hat\triangle^1)}$$

$$\mu_4 = \frac{\mu_{11}\triangle_l}{\hat\triangle^1 + \dfrac{\tau_l\mu_{11}\triangle_{l-1} + \hat\triangle^1}{\mu_{12}\triangle_{l-1} + \hat\triangle^1}\hat\triangle^2}$$

$$\mu_5 = \frac{\sigma_l\mu_3}{\hat\tau^1\hat\triangle^1}\left(\hat\triangle^1 + \frac{(\hat\epsilon^1 - \hat\tau^1)\triangle_{l-1}}{\hat\epsilon^1\triangle_{l-1} + \triangle_l}\hat\triangle^2\right)$$

$$\mu_6 = \hat\epsilon^1\hat\triangle^1\frac{\hat\epsilon^1\triangle_{l-1} + \triangle_l}{\triangle_l(\hat\gamma^1\triangle_{l-1} + \hat\triangle^1) - \hat\epsilon^1\triangle_{l-1}\hat\triangle^2}$$

$$\mu_7 = \frac{\triangle_l}{\dfrac{\delta_{l+1}}{\rho_{l+1}\mu_{13}}\dfrac{\rho_{l+1}\mu_{13}\hat\triangle^2 + \sigma_{l+1}\triangle_{l+1}}{\delta_{l+1}\mu_2\hat\triangle^2 + \epsilon_{l+1}\triangle_{l+1}}(\triangle_l - \mu_2\hat\triangle^2) + \hat\triangle^2}$$

$$\mu_8 = \frac{\mu_{13}\Delta_l}{\dfrac{\hat{\Delta}^2 + \rho_{l+1}\mu_{13}\Delta_{l+1}}{\hat{\Delta}^2 + \mu_{14}\Delta_{l+1}}\hat{\Delta}^1 + \hat{\Delta}^2}$$

$$\mu_9 = \frac{\sigma_{l+1}\mu_7}{\hat{\sigma}^2\hat{\Delta}^2}\left(\frac{(\hat{\delta}^2 - \hat{\rho}^2)\Delta_{l+1}}{\Delta_l + \hat{\delta}^2\Delta_{l+1}}\hat{\Delta}^1 + \hat{\Delta}^2\right)$$

$$\mu_{10} = \hat{\delta}^2\hat{\Delta}^2\frac{\Delta_l + \hat{\delta}^2\Delta_{l+1}}{\Delta_l(\hat{\Delta}^2 + \hat{\gamma}^2\Delta_{l+1}) - \hat{\delta}^2\hat{\Delta}^1\Delta_{l+1}}$$

$$\mu_{11} = \frac{\Delta_l}{\hat{\Delta}^1 + \dfrac{\tau_l\Delta_{l-1} + \hat{\Delta}^1}{\epsilon_l\Delta_{l-1} + \hat{\Delta}^1}\hat{\Delta}^2}$$

$$\mu_{12} = \frac{\hat{\epsilon}^1}{1 + (\hat{\gamma}^1 - \hat{\epsilon}^1)\dfrac{\Delta_{l-1}\hat{\Delta}^2}{(\hat{\gamma}^1\Delta_{l-1} + \Delta_l)\hat{\Delta}^1}}$$

$$\mu_{13} = \frac{\Delta_l}{\dfrac{\hat{\Delta}^2 + \rho_{l+1}\Delta_{l+1}}{\hat{\Delta}^2 + \delta_{l+1}\Delta_{l+1}}\hat{\Delta}^1 + \hat{\Delta}^2}$$

$$\mu_{14} = \frac{\hat{\delta}^2}{1 + (\hat{\gamma}^2 - \hat{\delta}^2)\dfrac{\hat{\Delta}^1\Delta_{l+1}}{(\Delta_l + \hat{\gamma}^2\Delta_{l+1})\hat{\Delta}^2}}$$

$$\mu_{15} = \frac{\Delta_l}{\hat{\Delta}^1 + \dfrac{\epsilon_l}{\tau_l\mu_{11}}\dfrac{\Lambda_l\Delta_{l-2} + \sigma_l\Delta_{l-1} + \tau_l\mu_{11}\hat{\Delta}^1}{\Theta_l\Delta_{l-2} + \delta_l\Delta_{l-1} + \epsilon_l\mu_1\hat{\Delta}^1}(\Delta_l - \mu_1\hat{\Delta}^1)}$$

$$\mu_{16} = \frac{\Delta_l}{\hat{\Delta}^1 + \dfrac{\epsilon_l}{\tau_l\mu_{11}}\dfrac{\tau_l(\Delta_l - \mu_{11}\hat{\Delta}^1) + \overline{\Lambda}_l\Delta_{l+1}}{\epsilon_l(\Delta_l - \mu_1\hat{\Delta}^1) + \overline{\Theta}_l\Delta_{l+1}}(\Delta_l - \mu_1\hat{\Delta}^1)}$$

$$\mu_{17} = \frac{\Delta_l}{\dfrac{\delta_{l+1}}{\rho_{l+1}\mu_{13}}\dfrac{\rho_{l+1}\mu_{13}\hat{\Delta}^2 + \sigma_{l+1}\Delta_{l+1} + \overline{\Phi}_{l+1}\Delta_{l+2}}{\delta_{l+1}\mu_2\hat{\Delta}^2 + \epsilon_{l+1}\Delta_{l+1} + \overline{\Theta}_{l+1}\Delta_{l+2}}(\Delta_l - \mu_2\hat{\Delta}^2) + \hat{\Delta}^2}$$

## References


[1] B. A. Barsky, *The $\beta$-spline: A local representation based on the shape parameters and on fundamental geometric measures*, Ph.D. thesis, University of Utah, Salt Lake City, UT, 1981.

[2] W. Boehm, *Curvature continuous curves and surfaces*, Comput. Aided Geom. Des., 2 (1985), pp. 313–323.

[3] ——, *Smooth curves and surfaces*, in Geometric Modeling: Algorithms and New Trends, G. Farin, ed., Society for Industrial and Applied Mathematics, Philadelphia, 1987, pp. 175–184.

[4] ——, *On the definition of geometric continuity*, Comput. Aided Des., 20 (1988), pp. 370–372.

[5] P. Dierckx and B. Tytgat, *Inserting new knots into beta-spline curves*, in Mathematical Methods in Computer Aided Geometric Design, T. Lyche and

L. L. Schumaker, eds., Academic Press, Boston, 1989.

[6] N. Dyn and C. A. Micchelli, *Piecewise polynomial spaces and geometric continuity of curves*, Tech. Rep., IBM Thomas J. Watson Research Center, Yorktown Heights, New York, 1985.

[7] M. Eck, *Allgemeine Konzepte geometrischer Bézier- und B-Spline-Kurven.* Masterthesis, Technische Hochschule, Darmstadt, Germany, 1987.

[8] M. Eck and D. Lasser, *B-spline-Bézier representation of geometric spline curves*, Tech. Rep., Technische Hochschule, Darmstadt, Germany, 1989.

[9] G. Farin, *Curves and Surfaces for Computer Aided Geometric Design*, Academic Press, 1988, Chap. 10.

[10] G. Geise, *Ueber beruehrende Kegelschnitte einer ebenen Kurve*, Z. Angew. Math. Mech., 42 (1962), pp. 297–304.

[11] T. N. T. Goodman, *Constructing piecewise rational curves with frenet frame continuity*, Comput. Aided Geom. Des., 7 (1990), pp. 15–31.

[12] T. N. T. Goodman and K. Unsworth, *Generation of $\beta$-spline curves using a recurrence relation*, in Fundamental Algorithm for Computer Graphics, Earnshaw, ed., NATO ASI Series, Springer-Verlag, New York, 1985, pp. 325–357.

[13] J. Hoschek and D. Lasser, *Grundlagen der geometrischen Datenverarbeitung*, Teubner, Stuttgart, 1989, Chap. 4.

[14] D. Lasser, *B-spline-Bézier representation of $vc^3$ and of $vc^4$ continuous spline curves*, Tech. Rep. # NPS-53-88-005, Naval Postgraduate School, Monterey, CA 93943, 1988.

[15] D. Lasser and M. Eck, *Bézier representation of geometric spline curves. A general concept and the quintic case*, Tech. Rep. # NPS-53-88-004, Naval Postgraduate School, Monterey, CA 93943, 1988.

[16] H. Pottmann, *Curves and tensor product surfaces with third order geometric continuity*, in Proceedings of the Third International Conference on Engineering Graphics and Descriptive Geometry, Vienna, Austria, 1988, pp. 107–116.

[17] ——, *Projectively invariant classes of geometric continuity for CAGD*, Comput. Aided Geom. Des., 6 (1989), pp. 307–322.

[18] P. Sablonniére, *Spline and Bézier polygons associated with a polynomial spline curve*, Comput. Aided Des., 10 (1978), pp. 257–261.

[19] H. P. Seidel, *Polynome, Splines und symmetrische, rekursive Algorithmen im CAGD.* Habilitationsschrift, Tübingen, 1989.

# On the Problem of Determining the Distance Between Parametric Curves

Frederick N. Fritsch and Gregory M. Nielson

## 7.1. Introduction

When comparing approximations to a given curve, it is necessary to have some comparison metric. In the functional case, a natural metric is the max-norm, which is the largest difference in $y$-values at corresponding points on the two curves:

$$\text{dist}(f,g) = \|f - g\|_\infty \approx \max_{0 \leq k \leq n} |f(x_k) - g(x_k)| \, ,$$

as illustrated in Fig. 7.1. This can be approximated arbitrarily well by taking a sufficiently large sample size $n$.

In the parametric case, it is natural to use Euclidean distance instead of a simple difference, but there is a problem in defining "corresponding points," due to possibly different parameterizations of the two curves.

The original motivation for this study was a desire to approximate Wilson-Fowler splines (WF-splines) [3] by nonuniform B-splines (NURBS). Any $G^2$ parametric piecewise cubic curve (e.g., a WF-spline) can be represented as a cubic B-spline curve with double knots [4]. Representing a WF-spline in this form requires a reparameterization. Consider the question, "How well can a

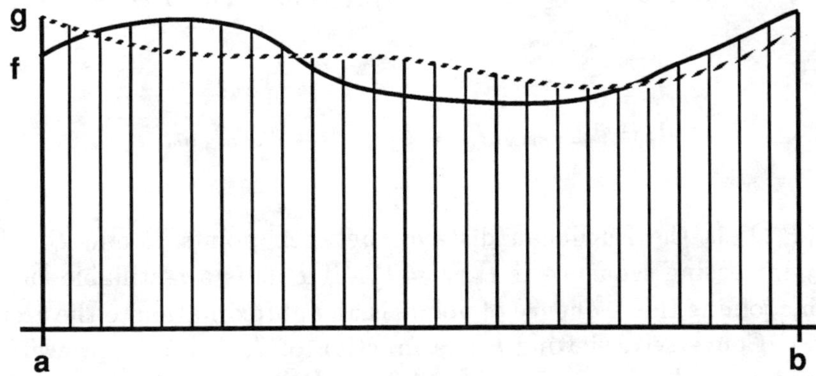


FIG. 7.1. *Comparison of two functions defined over* $[a,b]$.

123

WF-spline be approximated by a $C^2$ cubic B-spline?" What kind of metric should be used to compare the approximation with the original WF-spline? A practical metric can be based only on a finite sampling of the curve and should be independent of parameterization. This problem is discussed briefly by Rice in [6], and the examples in his Fig. 1 are very instructive. An approach that uses piecewise linear approximations to the two curves has been proposed by Emery [1], but this appears to be too specialized for application to general curves.

In §7.2 we discuss a practical comparison metric based on sampling the two curves at a finite number of points. Two variants, based on linear and cubic interpolation, respectively, are considered. In §7.3 the metric is successfully applied to two copies of the same curve with significantly different parameterizations, and convergence of the two variants is compared. In §7.4 examples are given to illustrate that our basic idea does not always give the intuitively correct answer. The final section discusses the question: "Where should we go from here?"

## 7.2.  A Proposed Curve Comparison Metric

**7.2.1.  The Main Idea.**   The proposed comparison metric is based on the fact that arclength is the canonical parameterization for a rectifiable curve. Let $N$ denote the number of sample points we are willing to use. Split $N = m + n$. The first $m$ evaluations will be used to obtain an approximation of arclength $s$ versus the curve parameter $t$. The remaining $n$ evaluations will be at points approximately uniformly spaced in arclength. These will be the corresponding points on the two curves for use in computing an approximate max-norm difference between the two curves. The optimal choice of $m$ and $n = N - m$ will, of course, be a compromise between getting a good approximation to $s(t)$ and obtaining a dense enough sample to get a good approximation to the distance between the curves.

**7.2.2.  Definition of the Metric.**   Let $P(t)$, $a \le t \le b$, represent one of the curves. Divide interval $[a, b]$ into $m$ subintervals of equal length $dt = (b-a)/m$. Let $t_j = a + j \cdot dt$ and evaluate $P_j = P(t_j)$, $j = 0, \cdots, m$. Define

$$(7.1) \qquad \begin{aligned} d_0 &= 0; \\ d_j &= d_{j-1} + \delta(P_{j-1}, P_j), \quad j = 1, \cdots, m, \end{aligned}$$

where $\delta(P, Q)$ is the Euclidean distance between points $P$ and $Q$. $\{d_j\}$ is a strictly increasing sequence if $P_{j-1} \ne P_j$. If $P(t)$ is a rectifiable curve, the broken line joining this sequence of points is an approximation to the arclength: $d_j \approx s(t_j)$. Conversely, viewing $t$ as a function of $d$, this is a piecewise linear approximation to the inverse function $\tau(d)$: $\tau(s(t)) = t$. Divide interval $[d_0, d_m]$ into $n$ subintervals of equal length $da = (d_m - d_0)/n$. Let $a_k = d_0 + k \cdot da$,

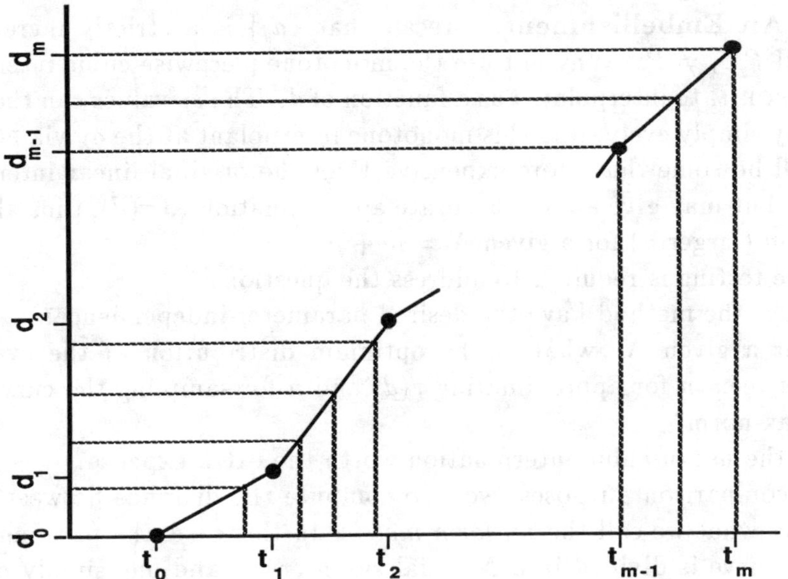



FIG. 7.2.  *Schematic representation of the proposed method.*

$k = 0, \cdots, n$. Compute

$$(7.2) \qquad s_k = t_j + (a_k - d_j)\frac{t_{j+1} - t_j}{d_{j+1} - d_j}, \quad k = 1, \cdots, n-1,$$

where $d_j \le a_k < d_{j+1}$ (linear interpolation in $t$ versus $d$ table). Note: $s_0 = t_0$, $s_n = t_m$.

This construction is illustrated in Fig. 7.2, where the dots are at $(t_j, d_j)$, $j = 0, \cdots, m$. The horizontal lines are uniformly spaced along the $d$-axis; the vertical lines indicate the locations of the $s_k$.

Finally, evaluate $\tilde{P}_k = P(s_k)$, $k = 0, \cdots, n$. These will be points approximately equally spaced in arclength along the curve. Because $s_0 = t_0$, $s_n = t_m$, we need only perform $n - 1$ new evaluations, for a total of $(m + 1) + (n - 1) = m + n$.

Follow this same procedure on the other curve to obtain $\{\tilde{Q}_k : k = 0, \cdots, n\}$, points approximately equally spaced in arclength along the curve $Q(t)$. The final comparison metric will be some vector norm applied to the sequence

$$(7.3) \qquad \delta_k = \delta(\tilde{P}_k, \tilde{Q}_k), \quad k = 0, \cdots, n.$$

We have chosen to use the max-norm:

$$(7.4) \qquad \mathrm{dist}(P, Q) \approx \max_{0 \le k \le n} \delta_k.$$

It is clear that this method can be used equally well for three- or higher-dimensional curves.

**7.2.3.  An Embellishment.**  Recall that $\{d_j\}$ is a strictly increasing sequence if $P_{j-1} \neq P_j$. Why not use the monotone piecewise cubic interpolation routine PCHIM to interpolate $t$ as a function of $d$? The $s_k$ values can then be obtained by simply evaluating this monotone interpolant at the $a_k$ via PCHFE [2]. This will be somewhat more expensive than the original linear interpolation method, but may give a more accurate approximation to $\tau(d)$, thus allowing a smaller $m$ (larger $n$) for a given $N = m + n$.

Some testing is required to address the questions:
- Does the method have the desired parameter-independence?
- For a given $N$, what is the optimum distribution of the evaluations between $m$ for approximating $\tau(d)$ and $n$ for sampling the curve for the max-norm?
- Is the use of cubic interpolation worth the extra expense?

For comparison purposes, we also compute the distance between the two curves by what we call the *uniform metric*. In this case, the parameter range for each curve is divided into $N$ equal subintervals and one simply compares the curves at the resulting points.

## 7.3.  Some Positive Examples

**7.3.1.  The Test Curves.**  The first tests were done on two different parameterizations of the same planar curve, for which the correct answer is zero. We considered two different parameterizations of a quarter-circle $P(t) = (R\cos u, R\sin u)$, $0 \leq t \leq 1$, where

$$(7.5) \qquad\qquad u = u_1(t) \equiv \frac{\pi}{2}t \,, \text{ or}$$

$$(7.6) \qquad\qquad u = u_2(t) \equiv \frac{\pi}{2}t^2 \,.$$

We also studied the following example due to Lee [5]:

$$(7.7) \qquad P(t) = \begin{cases} (0,1/3) + \phi(1-t)(-1,2/3), & 0 \leq t \leq 1; \\ (0,1/3) + \phi(t-1)(1,2/3), & 1 \leq t \leq 2. \end{cases}$$

Here $\phi$ is any increasing function in $C^2[0,1]$ with $\phi(0) = 0$, $\phi(1) = 1$. It is not hard to see that this curve consists of two straight line segments joining the three points $(-1,1)$ at $t = 0$, $(0,1/3)$ at $t = 1$, and $(1,1)$ at $t = 2$. If one chooses $\phi(u) = u^{k+1}$, $k > 0$, then this curve is $C^k$, even though its geometric continuity is clearly only zero. (The standard parameterization results from the choice $k = 0$.)

**7.3.2.  Behavior as $N$ Increases.**  Tables 7.1–7.4 contain summaries of some test runs using $m = n = N/2$. The linear approximation appears to converge quadratically. The PCHIM interpolant exhibits cubic convergence for Lee's example, but only quadratic for the quarter-circle example.

Figures 7.3–7.5 show different approximations to $s(t)$ for $k = 2$ with $N$ fixed, $m$ varying. The solid vertical lines indicate the locations of the $s_k$ for

TABLE 7.1

*Quarter-circle example.*

| N | Linear | (ratio) | Cubic | (ratio) | Uniform |
|---|---|---|---|---|---|
| 20 | 4.55746e-3 | | 7.85052e-4 | | 3.90181e-1 |
| 40 | 1.17315e-3 | (3.88) | 2.03104e-4 | (3.87) | 3.90181e-1 |
| 80 | 2.92580e-4 | (4.01) | 5.04631e-5 | (4.02) | 3.90181e-1 |
| 160 | 7.29056e-5 | (4.01) | 1.26700e-5 | (3.98) | 3.90181e-1 |
| 320 | 1.83873e-5 | (3.96) | 3.16161e-6 | (4.01) | 3.90181e-1 |
| 640 | 4.61432e-6 | (3.98) | 7.89396e-7 | (4.01) | 3.90181e-1 |

TABLE 7.2

*Lee's $C^k$ example, $k = 2$ versus $k = 0$.*

| N | Linear | (ratio) | Cubic | (ratio) | Uniform |
|---|---|---|---|---|---|
| 20 | 3.16044e-2 | | 2.06542e-2 | | 4.61511e-1 |
| 40 | 8.01206e-3 | (3.94) | 2.12025e-3 | (9.74) | 4.61511e-1 |
| 80 | 2.06295e-3 | (3.88) | 2.42353e-4 | (8.75) | 4.62581e-1 |
| 160 | 5.28032e-4 | (3.91) | 2.90172e-5 | (8.35) | 4.62581e-1 |
| 320 | 1.34583e-4 | (3.92) | 3.55117e-6 | (8.17) | 4.62581e-1 |
| 640 | 3.39805e-5 | (3.96) | 4.39261e-7 | (8.08) | 4.62591e-1 |
| 1280 | 8.59349e-6 | (3.95) | 5.46213e-8 | (8.04) | 4.62591e-1 |

TABLE 7.3

*Lee's $C^k$ example, $k = 3$ versus $k = 0$.*

| N | Linear | (ratio) | Cubic | (ratio) | Uniform |
|---|---|---|---|---|---|
| 20 | 5.37043e-2 | | 6.29210e-2 | | 5.65350e-1 |
| 40 | 1.57539e-2 | (3.41) | 5.95409e-3 | (10.57) | 5.66665e-1 |
| 80 | 4.25298e-3 | (3.70) | 6.57588e-4 | (9.05) | 5.67769e-1 |
| 160 | 1.09681e-3 | (3.88) | 7.74469e-5 | (8.49) | 5.67769e-1 |
| 320 | 2.78054e-4 | (3.94) | 9.40142e-6 | (8.24) | 5.67834e-1 |
| 640 | 6.99740e-5 | (3.97) | 1.15822e-6 | (8.12) | 5.67834e-1 |
| 1280 | 1.75498e-5 | (3.99) | 1.43733e-7 | (8.06) | 5.67839e-1 |

TABLE 7.4

*Lee's $C^k$ example, $k = 2$ versus $k = 3$.*

| N | Linear | (ratio) | Cubic | (ratio) | Uniform |
|---|---|---|---|---|---|
| 20 | 3.10927e-2 | | 4.22668e-2 | | 1.23670e-1 |
| 40 | 1.10630e-2 | (3.06) | 3.83383e-3 | (11.02) | 1.26758e-1 |
| 80 | 2.60249e-3 | (3.91) | 4.15235e-4 | (9.23) | 1.26758e-1 |
| 160 | 7.87999e-4 | (3.30) | 4.96186e-5 | (8.37) | 1.26758e-1 |
| 320 | 2.41155e-4 | (3.27) | 6.19683e-6 | (8.01) | 1.26758e-1 |
| 640 | 6.54676e-5 | (3.68) | 7.73892e-7 | (8.01) | 1.26758e-1 |
| 1280 | 1.69931e-5 | (3.85) | 9.66810e-8 | (8.00) | 1.26758e-1 |

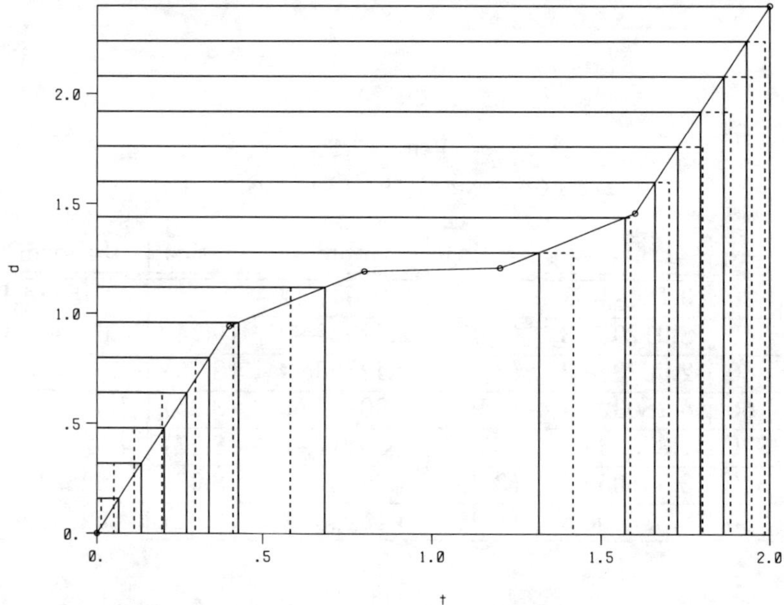

FIG. 7.3. *Plot of d versus t for Lee's $C^2$ example with $m = 5$, $n = 15$.*

the linear interpolant; the dashed lines, the cubic interpolant. Comparing Figs. 7.4 and 7.6 illustrates the convergence of the $d_j$ as $N$ increases. Figure 7.7 contains plots of the $k = 2$ and $k = 3$ curves, offset from one another so that lines can be drawn joining corresponding points used by the uniform metric. In Figs. 7.8–7.9 are given two comparable plots for the metric proposed here with $m = n$.

**7.3.3. Is There an Optimal $m$?**  We had expected that as we increased $m$, for fixed $N$, the results would get better for a while, then deteriorate, with an optimum value near $m = N/2$. To the contrary, preliminary results showed a steady decrease with $m$. Figure 7.10 contains the detailed output from one run with $N = 40$. Note that the odd and even subsequences behave quite differently. Each appears to be converging to zero as $m$ increases, with

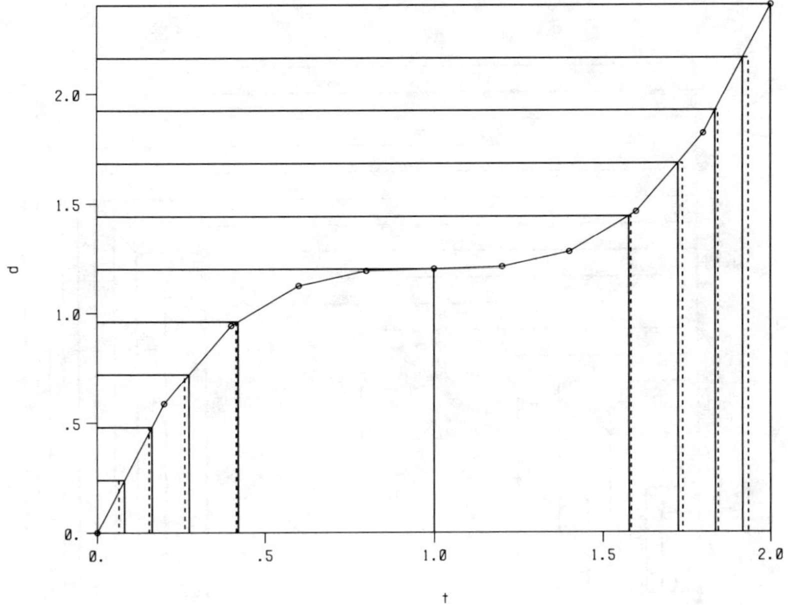

FIG. 7.4. *Plot of d versus t for Lee's $C^2$ example with $m = 10$, $n = 10$.*

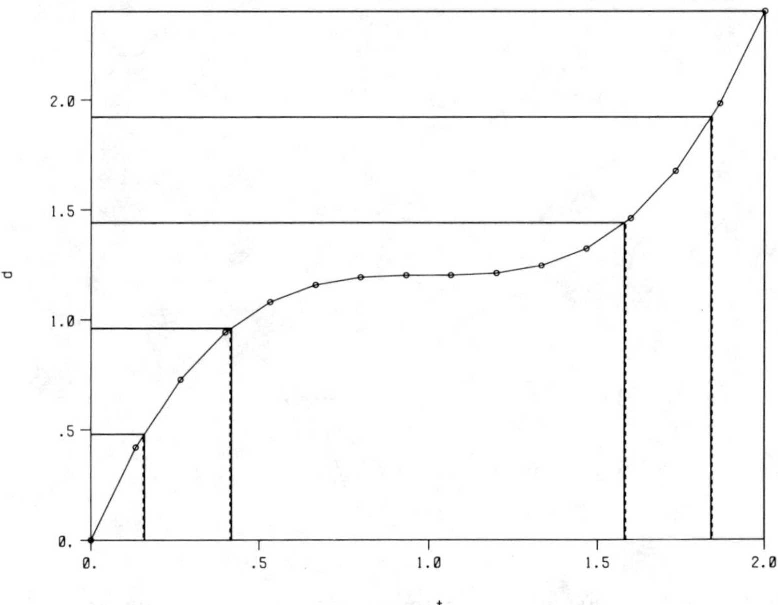

FIG. 7.5. *Plot of d versus t for Lee's $C^2$ example with $m = 15$, $n = 5$.*

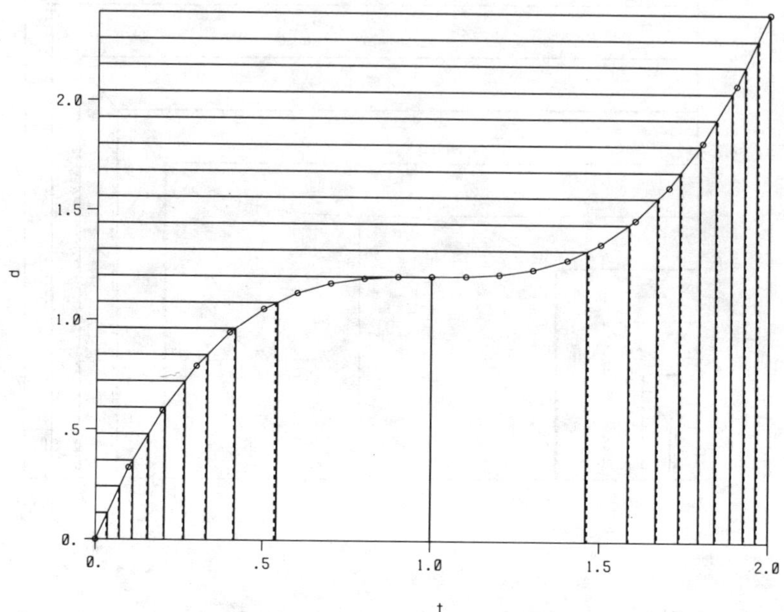

FIG. 7.6. *Plot of d versus t for Lee's $C^2$ example with $m = 20$, $n = 20$.*

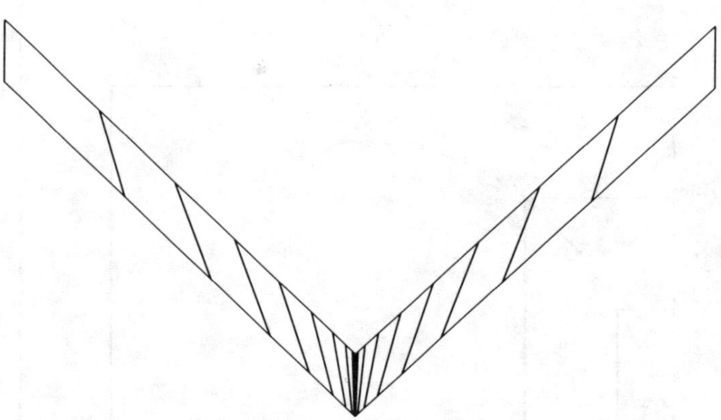

FIG. 7.7. *Uniform metric comparison points for Lee's $C^2$ (upper) and $C^3$ (lower) examples with $N = 20$.*

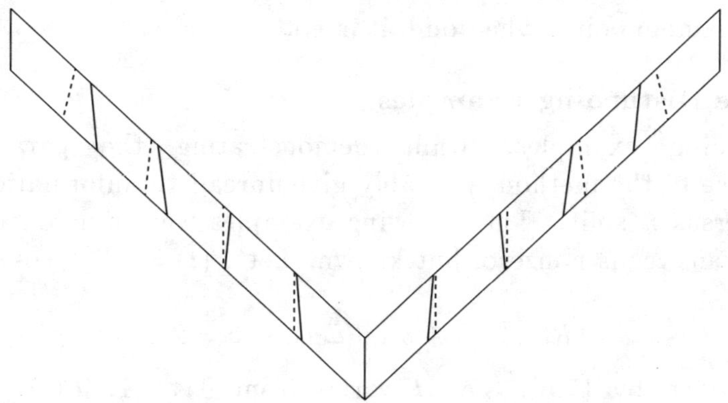

FIG. 7.8.   *Proposed metric comparison points for $C^2$ versus $C^3$ examples with $m = n = 10$. Solid lines are from linear method; dashed, cubic method.*

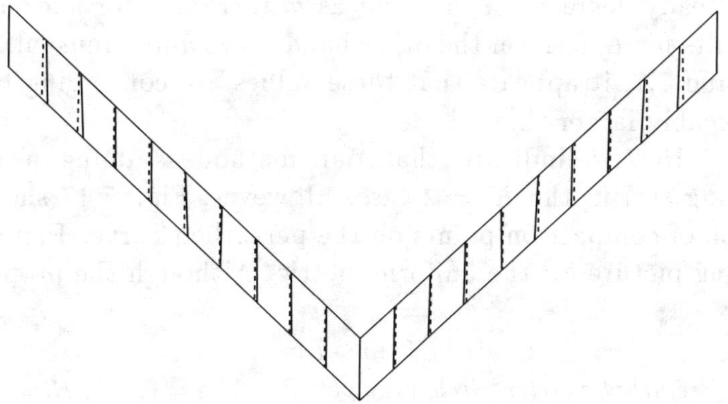

FIG. 7.9.   *Proposed metric comparison points for same with $m = n = 20$.*

the even sequence converging faster. The even subsequence for the cubic case converges faster than the linear case, as the above results indicate, but the odd subsequence looks very much like its linear counterpart. This was not resolved completely but we did make the following observation: when $m$ is even, the corner is a sample point; when odd, it is not.

## 7.4. Some Disturbing Examples

The preceding examples, while demonstrating the parameterization-independence of the method, probably give unrealistic information about optimal $m$ versus $n$ split. The following examples were studied next because the correct answer is nonzero, but known. Let $P(t) = (R(t)\cos u, R(t)\sin u)$, where

$$(7.8) \qquad R(t) = R + a\sin(\omega u), \quad \omega = 2K,$$

and $u$ is given by (7.5). As $t$ varies from 0 to 1, $P(t)$ traces out a perturbed quarter-circle of radius $R$. The perturbation has amplitude $a$ and has $K$ extrema over the interval. The unperturbed quarter-circle with parameterization (7.6) is used as the comparison curve. The "correct" distance between the curves, as illustrated in Fig. 7.11, is clearly

$$(7.9) \qquad \max_{0 \leq t \leq 1} [\, R(t) - R \,] = a.$$

The results for $K = 1$ were about what we had expected. These are summarized in Table 7.5 for the case $m = n = N/2$. The difference between $a$ and the computed distance is listed in the "error" column. Figure 7.12 shows corresponding points for the uniform metric. Figures 7.13 and 7.14 give comparable new metric points for two values of $N$. Interestingly, we still observed a steady decrease in the error as $m$ increases, for fixed $N = m + n$.

The results for $K > 1$, on the other hand, were quite unusual. Referring to Tables 7.6 and 7.7, it appears that these values are converging to some limit that is noticeably larger than $a$!

Figures 7.15–7.16 indicate that our method is doing a good job of approximating $s(t)$ in the $K = 2$ case. However, Fig. 7.17 shows a strange configuration of comparison points on the perturbed curve. Figure 7.18 is the corresponding picture for the uniform metric. Although the proposed method

TABLE 7.5

*Perturbed quarter-circle example, $R = 1$, $a = 0.123$, $K = 1$.*

| N | Linear | error | Cubic | error | Uniform |
|---|---|---|---|---|---|
| 20 | 0.123014 | -1.37e-5 | 0.123003 | -2.81e-6 | 0.431388 |
| 40 | 0.123002 | -2.01e-6 | 0.123000 | -1.78e-7 | 0.431388 |
| 80 | 0.123000 | -2.81e-7 | 0.123000 | -1.13e-8 | 0.431388 |
| 160 | 0.123000 | -2.43e-8 | 0.123000 | -7.28e-10 | 0.431388 |
| 320 | 0.123000 | -4.92e-10 | 0.123000 | -4.52e-11 | 0.431388 |

```
m= 2, n=38, l-metric=4.62587e-01, c-metric=4.62587e-01
m= 3, n=37, l-metric=2.95986e-01, c-metric=1.81926e-01
m= 4, n=36, l-metric=1.68984e-01, c-metric=1.79107e-01
m= 5, n=35, l-metric=1.30655e-01, c-metric=1.29687e-01
m= 6, n=34, l-metric=8.34438e-02, c-metric=1.18377e-01
m= 7, n=33, l-metric=7.16559e-02, c-metric=7.80379e-02
m= 8, n=32, l-metric=4.87815e-02, c-metric=4.95062e-02
m= 9, n=31, l-metric=4.52998e-02, c-metric=2.77577e-02
m=10, n=30, l-metric=3.20592e-02, c-metric=2.15233e-02
m=11, n=29, l-metric=3.22650e-02, c-metric=1.84502e-02
m=12, n=28, l-metric=2.29332e-02, c-metric=1.15551e-02
m=13, n=27, l-metric=2.43239e-02, c-metric=1.53775e-02
m=14, n=26, l-metric=1.66658e-02, c-metric=6.50667e-03
m=15, n=25, l-metric=1.93869e-02, c-metric=1.36549e-02
m=16, n=24, l-metric=1.31997e-02, c-metric=4.25791e-03
m=17, n=23, l-metric=1.65115e-02, c-metric=1.10598e-02
m=18, n=22, l-metric=9.95039e-03, c-metric=3.00763e-03
m=19, n=21, l-metric=1.41641e-02, c-metric=1.04292e-02
m=20, n=20, l-metric=8.01206e-03, c-metric=2.12025e-03
m=21, n=19, l-metric=1.19735e-02, c-metric=1.00212e-02
m=22, n=18, l-metric=7.04223e-03, c-metric=1.44440e-03
m=23, n=17, l-metric=8.85939e-03, c-metric=9.64401e-03
m=24, n=16, l-metric=5.94459e-03, c-metric=8.99022e-04
m=25, n=15, l-metric=9.83070e-03, c-metric=8.74422e-03
m=26, n=14, l-metric=4.54184e-03, c-metric=4.47871e-04
m=27, n=13, l-metric=8.33466e-03, c-metric=7.63511e-03
m=28, n=12, l-metric=2.76094e-03, c-metric=1.07551e-04
m=29, n=11, l-metric=6.84302e-03, c-metric=6.39812e-03
m=30, n=10, l-metric=3.11234e-03, c-metric=1.54016e-05
m=31, n= 9, l-metric=6.20923e-03, c-metric=5.78049e-03
m=32, n= 8, l-metric=3.18739e-03, c-metric=1.26237e-05
m=33, n= 7, l-metric=5.95486e-03, c-metric=5.23511e-03
m=34, n= 6, l-metric=1.49511e-03, c-metric=6.60562e-06
m=35, n= 5, l-metric=5.99224e-03, c-metric=4.60058e-03
m=36, n= 4, l-metric=1.76819e-03, c-metric=5.57993e-06
m=37, n= 3, l-metric=5.21107e-03, c-metric=3.62760e-03
m=38, n= 2, l-metric=3.44906e-14, c-metric=3.44906e-14
```

FIG. 7.10. *Variation of metric values with $m$ for Lee's $C^2$ versus $C^0$ example with $m + n = 40$.*

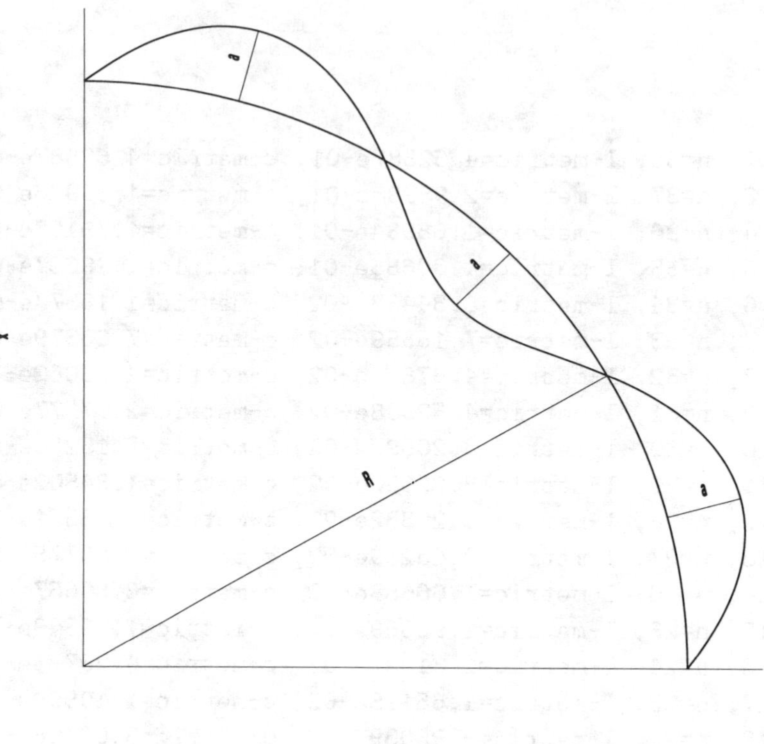



FIG. 7.11. *Distance between quarter-circle and a $K = 3$ perturbation.*

FIG. 7.12. *Uniform metric comparison points for perturbed quarter-circle example, $R = 1$, $a = 0.123$, $K = 1$, $N = 20$.*

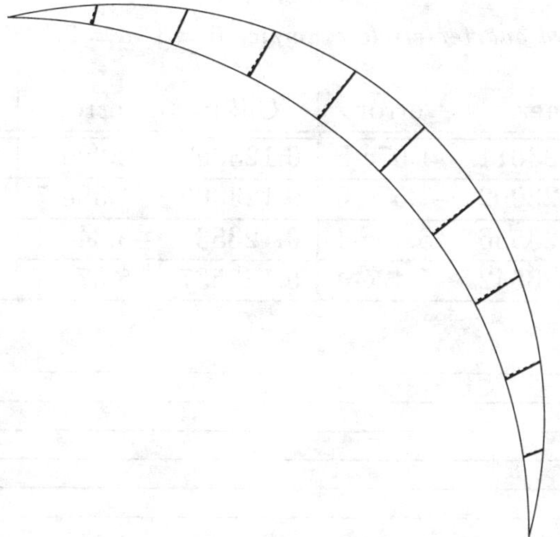

FIG. 7.13. *Proposed metric comparison points for perturbed quarter-circle with* $R = 1$, $a = 0.123$, $K = 1$, $m = n = 10$.

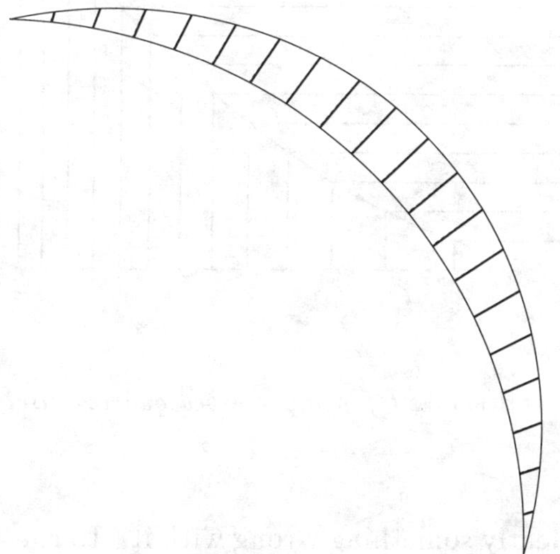

FIG. 7.14. *Proposed metric comparison points for same with* $m = n = 20$.

TABLE 7.6

*Perturbed quarter-circle example,* $R = 1$, $a = 0.123$, $K = 2$.

| N | Linear | error | Cubic | error | Uniform |
|---|---|---|---|---|---|
| 20 | 0.123632 | -6.32e-4 | 0.124578 | -1.58e-3 | 0.395433 |
| 40 | 0.125606 | -2.61e-3 | 0.125826 | -2.83e-3 | 0.395975 |
| 80 | 0.126551 | -3.55e-3 | 0.126557 | -3.56e-3 | 0.395975 |
| 160 | 0.126573 | -3.57e-3 | 0.126575 | -3.58e-3 | 0.395994 |

TABLE 7.7

*Perturbed quarter-circle example, $R = 1$, $a = 0.123$, $K = 3$.*

| N | Linear | error | Cubic | error | Uniform |
|---|--------|-------|-------|-------|---------|
| 20 | 0.123011 | -1.07e-5 | 0.123002 | -2.20e-6 | 0.358544 |
| 40 | 0.123002 | -1.57e-6 | 0.123000 | -1.39e-7 | 0.358544 |
| 80 | 0.123556 | -5.56e-4 | 0.123534 | -5.34e-4 | 0.358544 |
| 160 | 0.123542 | -5.42e-4 | 0.123537 | -5.37e-4 | 0.358544 |

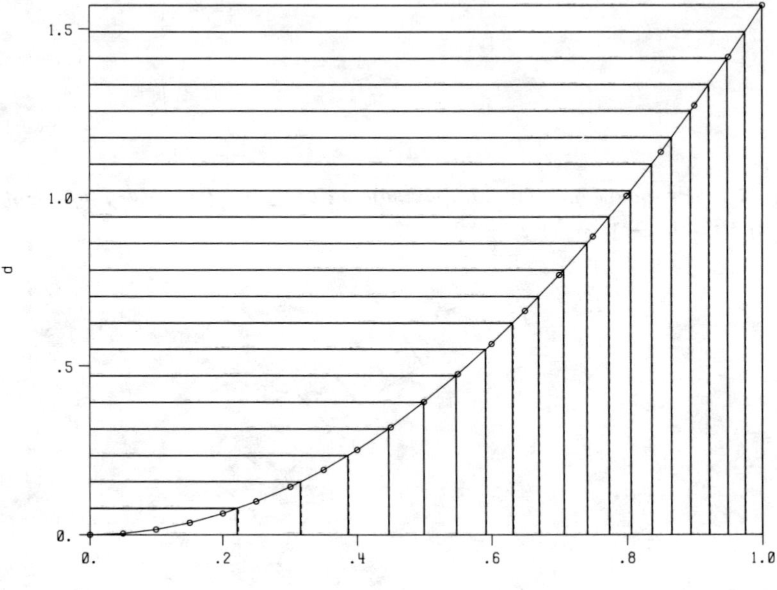

FIG. 7.15.  *Plot of d versus t for unperturbed quarter-circle with $m = n = 20$.*

is better, there is clearly something wrong with it. To check for possible coding errors, the comparison curve was rewritten with the standard parameterization (7.5), with very similar results (except that the uniform metric now rapidly converges to the correct answer). Figure 7.19 is the plot of the uniform metric comparison points for $N = 40$. It looks much more like what we had expected than Fig. 7.20, the results from our algorithm with $m = n = 40$. Note that the points on the quarter-circle are essentially identical for the two methods, but our method seems to be doing something wrong for the perturbed curve. Figures 7.21–7.22 are a similar pair for $K = 3$. Extensive tests and additional computations too detailed to describe here revealed the problem: there is more arclength in one of the lobes outside the circle than in an inner lobe. Thus, *comparing points uniformly spaced in arclength does <u>not</u> give the correct answer for these examples!* (The standard polar angle or something equivalent to it is needed here.)

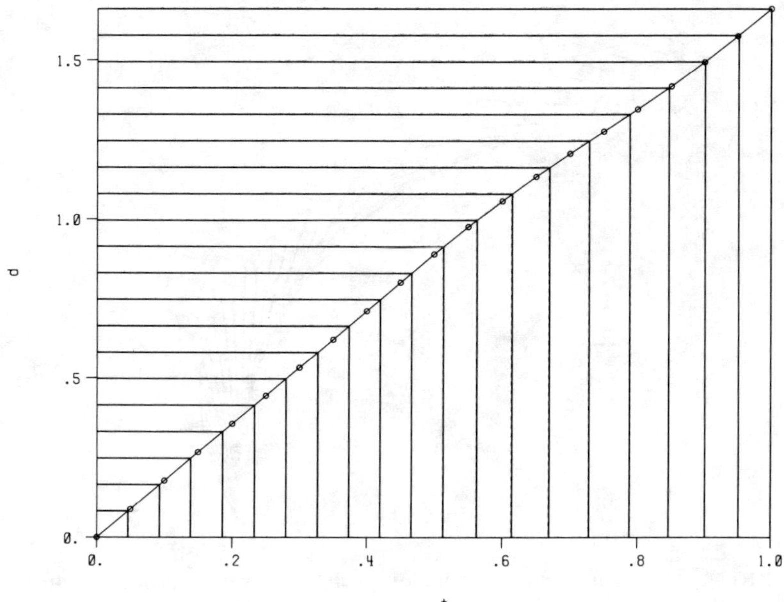

FIG. 7.16.  *Plot of d versus t for perturbed quarter-circle with $R = 1$, $a = 0.123$, $K = 2$, $m = n = 20$.*

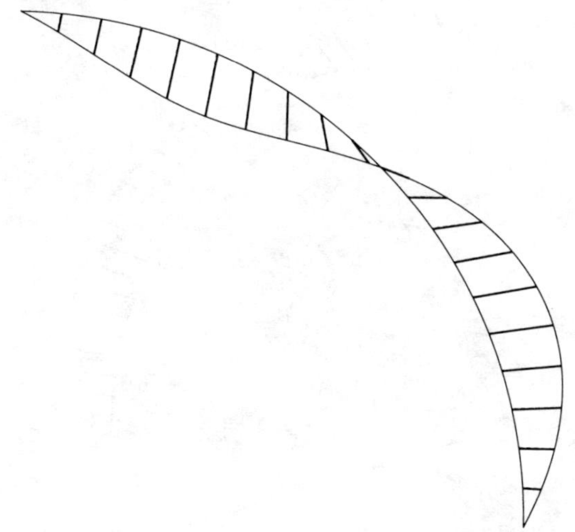

FIG. 7.17.  *Proposed metric comparison points, nonstandard quarter-circle versus perturbed quarter-circle with $R = 1$, $a = 0.123$, $K = 2$, $m = n = 20$.*

FIG. 7.18.   *Uniform comparison points for same with $N = 40$.*

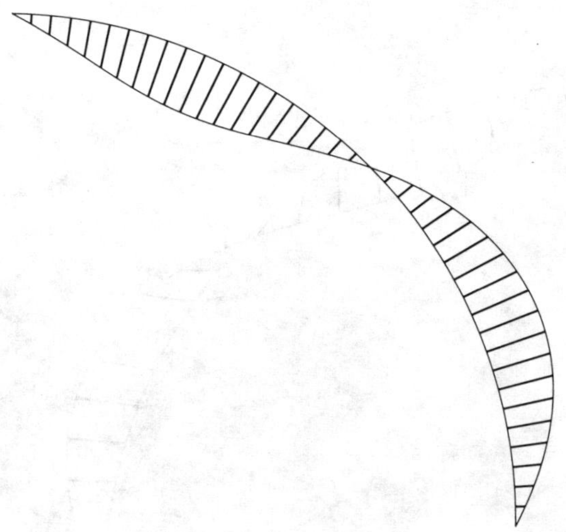

FIG. 7.19.   *Uniform metric comparison points, standard quarter-circle versus perturbed quarter-circle with $R = 1$, $a = 0.123$, $K = 2$, $N = 40$.*

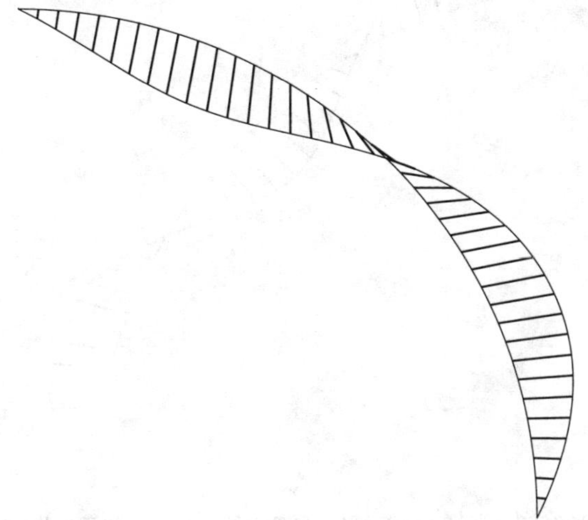

FIG. 7.20.  *Proposed metric comparison points for same with $m = n = 40$.*

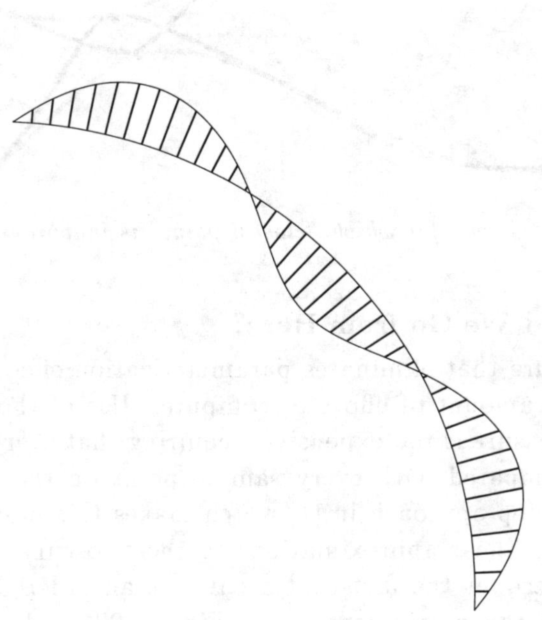

FIG. 7.21.  *Uniform metric comparison points, standard quarter-circle versus perturbed quarter-circle with $R = 1$, $a = 0.123$, $K = 3$, $N = 40$.*

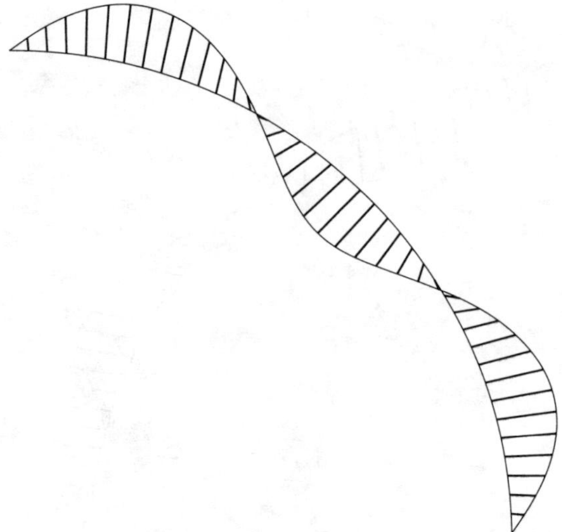

FIG. 7.22.  *Proposed metric comparison points for same with $m = n = 40$.*

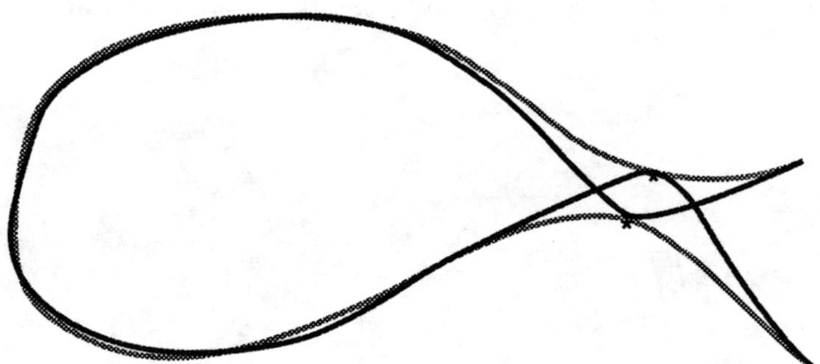

FIG. 7.23.  *Curves for which "closest point" is inappropriate.*

## 7.5.  Where Should We Go from Here?

We need some measure that eliminates parameterization effects and will not require a prohibitive amount of effort to compute.  Use of the standard set-theoretic distance measure is too expensive, requiring that every sample point on each curve be compared with every sample point on the other.  Emery has proposed a two-step approach in [1] which makes this more practical by introducing piecewise linear approximations to the two curves.  However, it does not take into account the fact that a curve is an ordered set of points.  For example, should the points marked in Fig.  7.23 really be considered "corresponding points"?

Suggestions for further work include

- Attempt to approximate the area between the curves, rather than using the max-norm.
- Use our method to obtain points on the standard curve and follow normals to the approximation.  (This requires solving $n + 1$ line-curve

intersection problems and still does not address the problem illustrated by Fig. 7.23.)

- Use the Bézier form with subdivision to obtain piecewise linear approximations for the two curves. (This applies only to piecewise polynomial curves, of course.)

All of these are much more complicated than our original proposal and contain numerous pitfalls.

## Acknowledgments

Some of the initial ideas of this research were formulated in a discussion between the authors and Tom Lyche during his visit to Lawrence Livermore National Laboratory in 1987. This work was performed under the auspices of the U. S. Department of Energy by Lawrence Livermore National Laboratory under contract W-7405-Eng-48. The second author was supported by DOE grant DE-FG02-87ER25041 to Arizona State University.

## References

[1] J. D. Emery, *The definition and computation of a metric on plane curves*, Comput. Aided Des., 18 (1986), pp. 25–28.

[2] F. N. Fritsch, *PCHIP Final Specifications*, LLNL Computer Documentation Report UCID-30194, 1982.

[3] ——, *History of the Wilson-Fowler Spline*, LLNL Informal Report UCID-20746, 1986.

[4] ——, *Representations for parametric cubic splines*, Comput. Aided Geom. Des., 6 (1989), pp. 79–82.

[5] E. T. Y. Lee, *Energy, Fairness, and a Counterexample*, private communication, 1988.

[6] J. R. Rice, *General purpose curve fitting*, in Approximation Theory, A. Talbot, ed., Academic Press, New York, 1970, pp. 191–204.

# Nontensor Product Surfaces

# A Survey of Parametric Scattered Data Fitting Using Triangular Interpolants

Stephen Mann, Charles Loop, Michael Lounsbery,
David Meyers, James Painter, Tony DeRose, and Kenneth Sloan

## 8.1. Introduction

The problem of passing a surface through a set of data points arises in numerous areas of application such as medical imaging, geological modeling, scientific visualization, and geometric modeling. Variants of this problem have been approached from many directions. Tensor-product B-splines work well for modeling surfaces based on rectilinear control nets but are not sufficient for more general topological types. Triangulated data, however, can represent arbitrary topological types. In this paper, we present a survey of a class of schemes that address the problem of fitting a surface to triangulated data.

One way to categorize surface fitting schemes is by the locality of data used in constructing a portion of the surface. A global scheme will use arbitrarily many of the data points in constructing each portion of the surface; a local scheme will only consider those points near the portion of the surface it is creating. Only local schemes are considered in this paper.

If the surface to be constructed lies above the plane it can be described as $\mathbf{S}(x,y) = (x,y,f(x,y))$. The data set is then referred to as scalar data, as the surface can be thought of as a scalar-valued function over the plane. Such data can be interpolated with a $C^1$ surface, using, for instance, the methods surveyed by Barnhill [1] and Franke [11].

A parametric scheme, on the other hand, constructs a vector-valued surface, $\mathbf{S}(u,v) = (x(u,v),y(u,v),z(u,v))$ and, unlike a scalar method, is capable of representing arbitrary topological types. The parametric problem is generally considered to be more difficult than the scalar variant. It has been shown, for instance, that the data cannot always be interpolated with a parametrically continuous surface [19]. Instead, the continuity conditions have to be relaxed to $G^1$- (tangent plane) continuity. The schemes surveyed in this paper are all parametric schemes.

In addition to geometric data, parametric interpolation schemes require information about the topological type of the desired surface. The topological information is usually specified as adjacency information relating the data points (vertices), edges, and faces. The schemes we have considered all assume that faces are triangular (i.e., three bounding edges per face), and that any number of faces may join at a vertex. These "triangular meshes" are sufficiently general to represent surfaces of arbitrary genus.

Finally, surface fitting schemes may interpolate or approximate the given data. Interpolating schemes construct surfaces that pass through the given data points. Approximating schemes produce surfaces that maintain the genus of the input data, but only pass near the data points. For some applications, an interpolating scheme is preferred, while for other applications, an approximating scheme may be a better choice. Here we will only consider interpolating schemes, and we will ignore the issue of specialized boundary conditions.

Thus, the primary goal of this paper is to present a unifying survey of local, parametric, triangular, interpolatory data fitting schemes. The surveyed schemes all proceed by first building boundary curves for a face and then filling in the interior of the face with one or more surface patches.

While all the schemes surveyed meet mathematical smoothness conditions, none of them produces surfaces with pleasing shape. Further, despite the diversity of the methods, all the schemes we implemented produced similar shape defects. Our investigations indicate that these poor shapes are primarily an artifact of the construction of boundary curves.

In §8.3, we present some background material. Three methods of constructing a tangent plane continuous join between two patches are presented in §8.4. In §8.5, the surveyed schemes are described. In §8.6, we look at the surfaces produced by these schemes and consider ways of improving their shapes. In §8.7, we summarize and present some recommendations.

## 8.2. Notation

Throughout this paper, scalars and scalar-valued functions will be denoted by *nonbold type letters* and Greek letters, such as $r$ and $\alpha$. Points and point-valued functions will be denoted with boldface letters, such as $\mathbf{V}$. Vectors will be represented by boldface letters topped with an arrow, such as $\vec{\mathbf{T}}$. Unit vectors will be denoted as $\hat{\mathbf{C}}$.

Surface patches will be denoted by the boldface letters $\mathbf{F}$ and $\mathbf{G}$. Usually, these surface patches will be in triangular Bézier form [10]. Often, we will consider the case when $\mathbf{F}$ and $\mathbf{G}$ are adjacent patches. In this case, the control points associated only with patch $\mathbf{F}$ will be denoted by $\mathbf{F}_i$, the control points associated only with patch $\mathbf{G}$ will be denoted by $\mathbf{G}_i$, and the control points common to both patches will be denoted by $\mathbf{H}_i$.

The vertices of a triangle in the domain of a patch will be denoted by $\mathbf{p}$, $\mathbf{q}$, and $\mathbf{r}$. The corresponding vertices in the range will be described by $\mathbf{V}_P$, $\mathbf{V}_Q$, and $\mathbf{V}_R$.

The directional derivative of a surface $\mathbf{F}$ in the direction $\vec{\mathbf{r}}$ is denoted by $\mathbf{D}_{\vec{r}}\mathbf{F}$. The derivative of a parametric curve $\mathbf{H}(t)$ is denoted $\mathbf{H}'(t)$. We will have use for a certain radial direction in the triangle $\mathbf{pqr}$, namely, $\vec{r}_{\mathbf{p}}(t) = ((1-t)\mathbf{q} + t\mathbf{r}) - \mathbf{p}$.

Finally, $B_i^n(t)$ will denote the $i$th Bernstein polynomial of $n$th degree, i.e.,

$$B_i^n(t) = \binom{n}{i}(1-t)^{n-i}t^i.$$

## 8.3. Background

Local interpolation schemes generally construct a surface consisting of multiple surface patches. In order for the entire surface to look smooth, certain continuity conditions must be met at every boundary between two patches. To avoid holes in the surface, every pair of neighboring patches must meet with $C^0$-continuity. To ensure that adjacent patches meet smoothly, one might also want them to meet with a continuous first derivative. However, this is not possible for surfaces of arbitrary genus. An alternate approach is to construct the surface patches to meet with continuous tangent planes along the boundaries. The patches are then said to meet with $G^1$-continuity (cf. [2], [20]). Several methods of ensuring tangent plane continuity will be presented in §8.4.

A second issue is what is sometimes referred to as the *vertex consistency problem*. This problem occurs when trying to construct a single $C^2$ patch for each triangular face of the data. The $G^1$-continuity conditions between patches set up a system of constraints around each data point. For a vertex of even degree greater than four, it has been shown that this system can not necessarily be satisfied [32].

There are primarily two approaches taken to avoid this problem. The first approach constructs multiple patches per face, which successfully decouples the cycle of constraints. A second approach is to construct patches that are not $C^2$ at the data points. The schemes surveyed in this paper all use one of these two approaches.

A third approach to solving the vertex consistency problem is to construct a "$C^2$ consistent" curve network. Peters has shown that if the boundary curves adjacent to a data point all agree with a common second fundamental form, then the above-mentioned cycle of constraints can be satisfied [28]. Note that while this is a sufficient condition for satisfying the vertex consistency problem, it is not a necessary condition. An alternate approach to constructing a $C^2$ consistent curve network can be found in [23].

## 8.4. Tangent Plane Continuity

Fitting surface patches together with tangent plane continuity has been approached from several directions. Farin [7] and Piper [29] give sufficient conditions for two polynomial patches to meet with $G^1$-continuity. A second approach, taken in [4], [19], [21], [26], is to first create a cross-boundary tangent

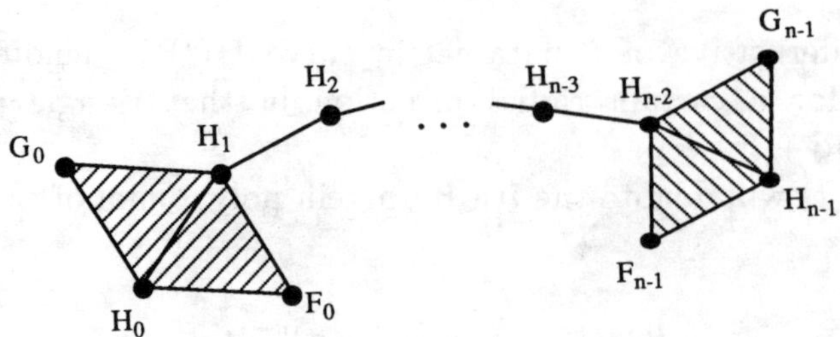

FIG. 8.1. *Bézier control points used in Farin's $G^1$ conditions.*

vector field for each boundary and then to construct patches that agree with these cross-boundary fields.

**8.4.1. Farin.** Farin [7] presented conditions for two degree $n$ polynomial patches with a common degree $n - 1$ boundary to meet with $G^1$-continuity. Labeling the Bézier control points as in Fig. 8.1, Farin's conditions are as follows.

Given two degree $n$ polynomial patches with a common degree $n - 1$ boundary, where

$$(8.1) \quad \begin{aligned} \mathbf{G}_0 &= \alpha_1 \mathbf{H}_0 + \alpha_2 \mathbf{H}_1 + \alpha \mathbf{F}_0, & \alpha_1 + \alpha_2 + \alpha = 1, \\ \mathbf{G}_{n-1} &= \alpha_3 \mathbf{H}_{n-2} + \alpha_4 \mathbf{H}_{n-1} + \alpha \mathbf{F}_{n-1}, & \alpha_3 + \alpha_4 + \alpha = 1, \end{aligned}$$

then the two patches meet with $G^1$-continuity if

$$\mathbf{G}_i = \frac{n - 1 - i}{n - 1}(\alpha_1 \mathbf{H}_i + \alpha_2 \mathbf{H}_{i+1} + \alpha \mathbf{F}_i) + \frac{i}{n - 1}(\alpha_3 \mathbf{H}_{i-1} + \alpha_4 \mathbf{H}_i + \alpha \mathbf{F}_i).$$

Equation (8.1) can be formulated in terms of the areas of the shaded triangles of Fig. 8.1:

$$\frac{\text{area}(\mathbf{G}_0, \mathbf{H}_0, \mathbf{H}_1)}{\text{area}(\mathbf{F}_0, \mathbf{H}_0, \mathbf{H}_1)} = \frac{\text{area}(\mathbf{G}_{n-1}, \mathbf{H}_{n-1}, \mathbf{H}_{n-2})}{\text{area}(\mathbf{F}_{n-1}, \mathbf{H}_{n-1}, \mathbf{H}_{n-2})}.$$

**8.4.2. Piper.** Piper [29] develops sufficient conditions for two quartic patches with quartic boundaries to meet with $G^1$-continuity. He begins by noting that the following equation must hold for patches $\mathbf{F}$ and $\mathbf{G}$ to meet $G^1$:

$$(8.2) \quad e(t)\mathbf{I}(t) + f(t)\mathbf{J}(t) + g(t)\mathbf{K}(t) + h(t)\mathbf{L}(t) = \vec{\mathbf{0}},$$

where $e, f, g,$ and $h$ are scalar functions such that

$$e(t) + f(t) + g(t) + h(t) = 0$$

for $t \in [0, 1]$, and where

$$\mathbf{I}(t) = \sum B_i^3(t)\mathbf{H}_{i+1},$$
$$\mathbf{J}(t) = \sum B_i^3(t)\mathbf{H}_i,$$

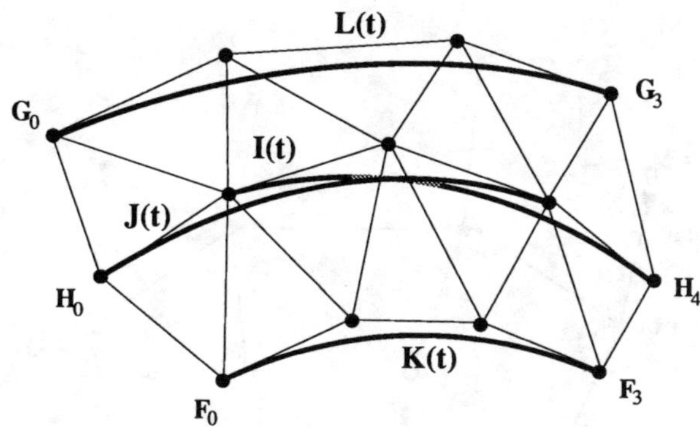



FIG. 8.2. *The curves used in Piper's $G^1$ conditions.*

$$\mathbf{K}(t) = \sum B_i^3(t)\mathbf{F}_i,$$

$$\mathbf{L}(t) = \sum B_i^3(t)\mathbf{G}_i.$$

These curves are illustrated in Fig. 8.2. Here, the points $\mathbf{F}_0$, $\mathbf{F}_3$, $\mathbf{G}_0$, $\mathbf{G}_3$, $\mathbf{H}_0$, $\mathbf{H}_1$, $\mathbf{H}_3$, and $\mathbf{H}_4$ are known.

Piper restricts $e$, $f$, $g$, and $h$ to be linear functions. Equation (8.2) then reduces to a $3 \times 5$ system of linear equations, where the unknowns are the control points $\mathbf{F}_1$, $\mathbf{F}_2$, $\mathbf{G}_1$, $\mathbf{G}_2$, and $\mathbf{H}_2$. In certain situations, the functions $e$, $f$, $g$, and $h$ are all constant functions. In this case, the $3 \times 5$ system of equations reduces to a $2 \times 5$ system. In summary, these constraints represent underdetermined conditions on the unknown control points to achieve a $G^1$ join of the patches.

### 8.4.3. Chiyokura-Kimura, Herron, Jensen.

As mentioned earlier, one approach to creating surface patches that meet with $G^1$-continuity is to first construct a field of cross-boundary tangent vectors along the boundary between two patches, using data common to both patches. The cross-boundary tangent field, together with the first derivative vector of the boundary curve, defines a tangent plane field all along the boundary. Next, the two patches are constructed, one on either side of the boundary, that match this tangent plane field along the boundary. The two patches will therefore meet with $G^1$-continuity. This is the approach taken in [4], [19], [21], [26]. We present now the method of Chiyokura and Kimura and show its relationship to other constructions.

Although Chiyokura and Kimura's cross-boundary construction was originally intended for rectangular patches, it readily extends to the construction of quartic triangular Bézier patches [31]. The method uses the boundary data for two adjacent patches to construct the Bézier control points that influence the tangent plane behavior along their common boundary. The boundary data consist of a cubic polynomial boundary curve and a pair of tangent vectors at each end of the boundary curve (Fig. 8.3 shows this data in Bézier form). Two quartic interior Bézier control points for each patch are set so as to match the

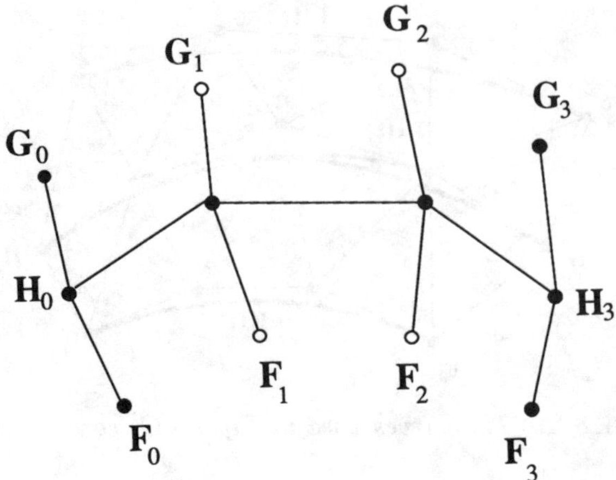



FIG. 8.3. *Boundary data used by Chiyokura and Kimura's construction. The solid points are the known control points; the hollow points are constructed so that the two patches meet $G^1$. Note that the **H**s are cubic control points, while the **F**s and **G**s are quartic control points.*

tangent plane field. We will give the construction for only one patch since the construction for the other patch is identical.

The cross-boundary tangent vector field is defined by linearly blending two vectors, one in each of the tangent planes at the end points. Chiyokura and Kimura choose these vectors to be unit vectors perpendicular to the tangents at the end points of the boundary curve. For patch $\mathbf{F}$, this blend is given by

$$\vec{\mathbf{C}}(t) = (1 - t)\hat{\mathbf{C}}_0 + t\hat{\mathbf{C}}_1.$$

The two perpendiculars $\hat{\mathbf{C}}_0$ and $\hat{\mathbf{C}}_1$ are unique, up to sign. The signs are chosen based on the vectors $\vec{\mathbf{F}}_0$ and $\vec{\mathbf{F}}_3$ as shown below, where $\vec{\mathbf{F}}_i = \mathbf{F}_i - \mathbf{H}_i$. $\vec{\mathbf{C}}(t)$ together with $\mathbf{H}'(t)$ completely specify the tangent plane field along the boundary.

For $\mathbf{F}$ to agree with the tangent plane field given by $\mathbf{H}'(t)$ and $\vec{\mathbf{C}}(t)$, there must exist functions $k(t)$ and $h(t)$ such that

$$(8.3) \qquad \mathbf{D}_{\vec{\mathbf{r}}(t)}\mathbf{F}(0, t, 1 - t) = k(t) \cdot \vec{\mathbf{C}}(t) + h(t) \cdot \mathbf{H}'(t),$$

where $\vec{\mathbf{r}}(t)$ is the radial direction in the domain of $\mathbf{F}$.

The values of $k(t)$ and $h(t)$ can be determined at the end points by evaluating (8.3) at $t = 0$ and $t = 1$:

$$\vec{\mathbf{F}}_0 = k_0 \cdot \hat{\mathbf{C}}_0 + h_0 \cdot \vec{\mathbf{H}}_0,$$

$$\vec{\mathbf{F}}_3 = k_1 \cdot \hat{\mathbf{C}}_1 + h_1 \cdot \vec{\mathbf{H}}_2,$$

where $\vec{\mathbf{H}}_i = \mathbf{H}_{i+1} - \mathbf{H}_i$, $k_0 = k(0)$, $k_1 = k(1)$, $h_0 = h(0)$, and $h_1 = h(1)$. For $h$ and $k$ to interpolate these end point conditions, they both must be at least linear functions. If we restrict them to be no more than linear, then each is uniquely determined:

$$k(t) = k_0 \cdot (1 - t) + k_1 \cdot t,$$

$$h(t) = h_0 \cdot (1 - t) + h_1 \cdot t.$$

Rewriting (8.3) in the cubic Bernstein basis, we can use the coefficients to $B_1^3(t)$ and $B_2^3(t)$ to determine the desired interior control points, resulting in

$$(8.4) \qquad \mathbf{F}_1 = \frac{1}{3}\{(k_0 + k_1)\hat{\mathbf{C}}_0 + k_0\hat{\mathbf{C}}_1 + 2h_0\vec{\mathbf{H}}_1 + h_1\vec{\mathbf{H}}_0\} + \mathbf{H}_1,$$

$$(8.5) \qquad \mathbf{F}_2 = \frac{1}{3}\{k_1\hat{\mathbf{C}}_0 + (k_0 + k_1))\hat{\mathbf{C}}_1 + h_0\vec{\mathbf{H}}_2 + 2h_1\vec{\mathbf{H}}_1\} + \mathbf{H}_2.$$

There is still some freedom left in (8.3). If $h(t)$ is a linear function, then the product $k(t) \cdot \vec{\mathbf{C}}(t)$ must be a polynomial of no higher than cubic degree. In the above formulation, this product is only a quadratic polynomial. Either $k(t)$ or $\vec{\mathbf{C}}(t)$ could be increased from a linear function to a quadratic function. Increasing $k(t)$ to a quadratic function gives a scalar degree of freedom, while increasing the degree of $\vec{\mathbf{C}}(t)$ yields a vector degree of freedom. Jensen [21] used the former generalization. He used the same linear blend of unit vectors for $\vec{\mathbf{C}}(t)$, but used the following quadratic scale function:

$$k^*(t) = k_0 \cdot u_0(t) + C \cdot \frac{(k_0 + k_1)}{2}u_1(t) + k_1 \cdot u_2(t),$$

where

$$u_0(t) = 2t^2 - 3t + 1, \quad u_1(t) = 4t - 4t^2, \quad u_2(t) = 2t^2 - t,$$

and $C$ is a scalar shape parameter.[1] For $C = 1$, $k^*(t) = k(t)$.

A second degree of freedom in (8.3) is in the choice of $\hat{\mathbf{C}}_0$ and $\hat{\mathbf{C}}_1$. These two vectors may be chosen in any fashion that uses information available to both patches, where the construction from both sides gives the same vectors with opposite sign. For example, in a later paper [3], Chiyokura defines $\hat{\mathbf{C}}_0$ and $\hat{\mathbf{C}}_1$ as

$$\hat{\mathbf{C}}_0 = \frac{\mathbf{G}_0 - \mathbf{F}_0}{|\mathbf{G}_0 - \mathbf{F}_0|},$$

$$\hat{\mathbf{C}}_1 = \frac{\mathbf{G}_3 - \mathbf{F}_3}{|\mathbf{G}_3 - \mathbf{F}_3|}.$$

Note that this definition of $\hat{\mathbf{C}}_0$ and $\hat{\mathbf{C}}_1$ is affine invariant and requires knowledge about both patches neighboring the boundary, whereas the earlier definition is not affine invariant and only uses information about the boundary curve.

Although Herron [19] approaches the problem somewhat differently, his construction and the Chiyokura-Kimura construction build the same field of cross-boundary tangent vectors along the boundary curves.

---

[1] In Jensen's paper, $u_1$ is given as $u_1(t) = 4t - t^2$. However, without the factor of 4 scaling $t^2$, $k^*(1)$ does not interpolate $k_1$.

## 8.5.  Parametric Schemes

The schemes studied in this survey fall into two categories: the split domain schemes and the convex combination schemes. The split domain schemes surveyed were presented in [21], [29], [31]. The convex combination schemes surveyed were presented in [19], [22], [26]. All methods surveyed build a surface for each triangular face by first computing boundary curves around the triangle, and then constructing one or more patches that match this boundary data. The two categories differ in their solution to the vertex consistency problem.

### 8.5.1.  Split Domain Schemes.

An extensive body of literature exists that discusses the properties of polynomial Bézier patches (cf. Farin [10]). If the data could be fit with Bézier patches, we could draw on this body of knowledge to compute various properties of the surface. However, if the boundary curves are constructed independently of each other, then a single Bézier patch cannot in general be used to interpolate the data, as the vertex consistency problem cannot in general be solved. Split domain schemes avoid this problem by constructing three patches per face, essentially splitting the domain triangle into three subtriangles, as was done by Clough and Tocher for scalar-valued data [9].

After splitting, each of the subpatches is used to interpolate the data along one of the boundaries. Splitting allows the data along each boundary to be matched independently of the data on the other two boundaries. The remaining degrees of freedom are used to make the internal boundaries of the three subpatches meet with $G^1$-continuity.

All three of the split domain schemes presented here construct quartic polynomial patches. Figure 8.4 schematically shows a labeling of the Bézier control points of these patches. Some of the schemes in this section compute cubic boundaries, so we will need to refer to both the cubic control points and the quartic control points for these boundaries. Symbols such as $\mathbf{I}^4$ and $\mathbf{E}^4$ refer to the quartic control points, whereas $\mathbf{I}^3$ and $\mathbf{E}^3$ refer to the corresponding cubic control points. The appearance of the resulting formulas is unfortunately visually complex; we have chosen it so that we can be specific about which quantities are to be used in the calculations.

**Shirman-Séquin.**  The following split domain scheme was proposed by Shirman and Séquin [31]. Three quartic triangular patches are constructed per face so as to interpolate the data points. The construction assumes that cubic boundary curves have been constructed and subsequently degree raised [10] to quartics.

Farin's $G^1$ conditions are used on the internal boundaries to ensure that the three patches meet each other with $G^1$-continuity. For $ijk \in \{PQR, QRP, RPQ\}$, the following relationships between the points $\mathbf{I}^3_{i1}$, $\mathbf{V}_i$, $\mathbf{E}^3_{i1}$, and $\mathbf{E}^3_{i2}$ are thus imposed:

$$(8.6) \qquad \mathbf{E}^4_{j3} = \alpha_{i2}\mathbf{I}^3_{i1} + \alpha_{i1}\mathbf{V}_i + \alpha_i\mathbf{E}^4_{k1},$$

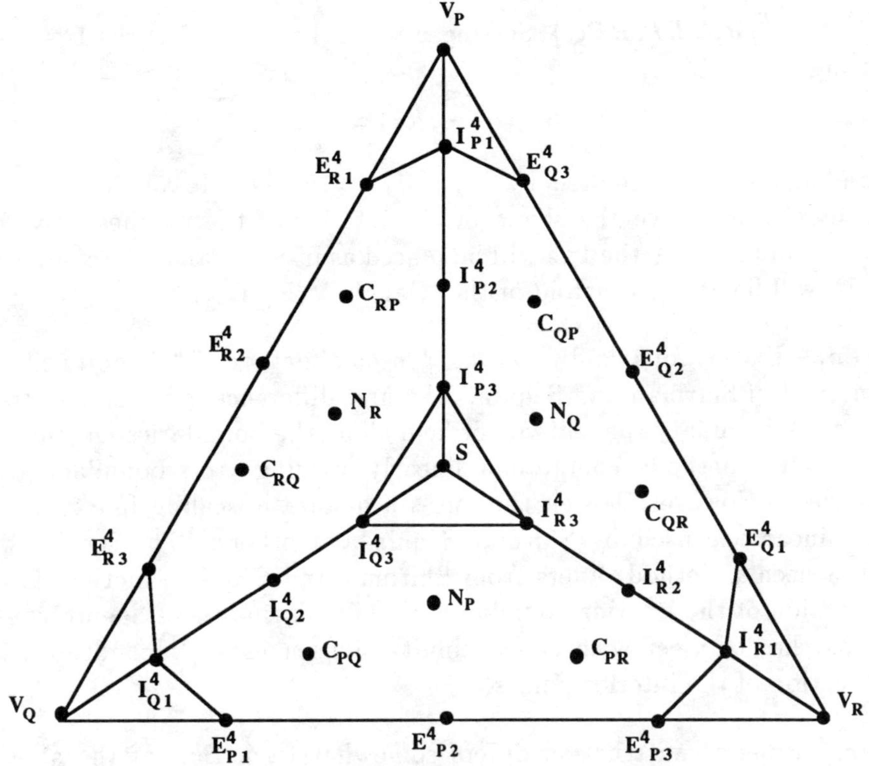



FIG. 8.4. *Bézier control points for split domain schemes.*

where $\alpha_{i1} + \alpha_{i2} + \alpha_i = 1$. Similarly, at the other end of the internal boundaries, the following relationships hold:

$$(8.7) \qquad \mathbf{I}^4_{k4} = \alpha_{i3}\mathbf{I}^3_{i3} + \alpha_{i4}\mathbf{S} + \alpha_i\mathbf{I}^4_{j4},$$

where $\alpha_{i3} + \alpha_{i4} + \alpha_i = 1$. For reasons of symmetry, we set $\alpha_i = -1$ for all $i$.

A setting of the $\alpha$s determines the control points $\mathbf{I}^3_{i1}$ according to (8.6). The method of Chiyokura and Kimura can now be used to establish tangent plane continuity across the external boundaries ((8.4) and (8.5)), thus determining six of the interior control points (the $\mathbf{C}_{ij}$s).

Using these $\alpha$s, Farin's continuity conditions set up a system of equations involving the interior control points which has the following solution:[2]

$$\mathbf{I}^3_{i2} = -\frac{\alpha_{i3}}{2\alpha_{i2}}\mathbf{V}_i - \left(\frac{\alpha_{i1}}{\alpha_{i2}} + \frac{\alpha_{i4}}{2\alpha_{i2}}\right)\mathbf{I}^3_{i1} + \frac{3}{2\alpha_{i2}}(\mathbf{C}_{ji} + \mathbf{C}_{ki}),$$

$$\mathbf{N}_i = -\frac{\alpha_{i3}}{3}\mathbf{I}^3_{i1} + \frac{\alpha_{i3}}{3}(\mathbf{I}^3_{j1} + \mathbf{I}^3_{k1}) + \left(\frac{\alpha_{i2}}{18} - \frac{\alpha_{i1} + 2\alpha_{i4}}{6}\right)\mathbf{I}^3_{i2} + \left(\frac{\alpha_{i2}}{18} + \frac{\alpha_{i1} + 2\alpha_{i4}}{6}\right)\left(\mathbf{I}^3_{k2} + \mathbf{I}^3_{j2}\right),$$

---

[2] An error was made in the published version of these equations. Here we present a correct solution.

for $ijk \in \{PQR, QRP, RPQ\}$. Setting $\mathbf{S}$ to be the centroid of the $\mathbf{I}_{i2}^3$s fixes the following $\alpha$s:

$$\alpha_{i3} = -\frac{3}{4}, \quad \alpha_{i4} = \frac{11}{4}.$$

$\alpha_{i1}$ and $\alpha_{i2}$ are now related by $\alpha_{i1} + \alpha_{i2} = 2$. This leaves a scalar shape parameter to influence the shape of the interior of the patches. By setting $\alpha_{i1} = -\frac{1}{4}$ and $\alpha_{i2} = \frac{9}{4}$ the $\mathbf{I}_{i1}^3$s will be placed as in Shirman and Séquin's paper (i.e., $\mathbf{I}_{i1}^3$ will lie at the centroid of the triangle $\mathbf{V}_i \mathbf{E}_{j1}^3 \mathbf{E}_{k3}^3$).

**Jensen.** Except for two differences, Jensen's method [21] is quite similar to the method of Shirman and Séquin. The first difference is in the construction of the cross-boundary tangent vector field along the boundaries. As mentioned earlier, both methods compute a linearly varying cross-boundary tangent vector field. However, Jensen then uses a quadratic scaling function instead of the linear one used by Shirman-Séquin and others. The second way in which Jensen's method differs from Shirman and Séquin's method is in the construction of the interior boundaries. While Shirman and Séquin construct their patches to meet with $G^1$-continuity, Jensen uses $C^1$ conditions in the construction of the interior points.

**Piper.** Piper's construction differs somewhat from that of the above two split domain schemes. First, a single cubic patch is constructed for each face. Next, this cubic patch is subdivided at the centroid into three cubic subpatches. These patches are then modified to produce "candidate" patches, which are degree elevated to quartic patches. The control points of the quartic patches are adjusted so that they satisfy the tangent plane continuity conditions of §8.4.2 along the exterior boundaries. As there are more than one set of such control points that satisfy Piper's continuity conditions (due to rank deficiency of the linear system), the set chosen is the one closest to the candidate control points in a least squares sense. Finally, the control points along the interior boundaries are adjusted so that the three patches meet each other with $C^1$-continuity.

**8.5.2. Convex Combination Schemes.** Convex combination schemes create a single patch for each face. The patches are $C^2$ everywhere except at the vertices. This successfully avoids the vertex consistency problem by not having consistently defined mixed partial (i.e., *twist*) terms at the patch corners. Each patch is constructed by first building boundary curves and tangent plane fields along these boundary curves. Next, three patches are created, each of which interpolates part of the boundary data. Finally, a single patch is formed by taking a convex combination of the three patches in such a way that the resulting patch interpolates all of the boundary data.

**Nielson.** Nielson [26] presented two surface construction techniques. The first is a transfinite method. The input to this scheme is a "triangle" of three

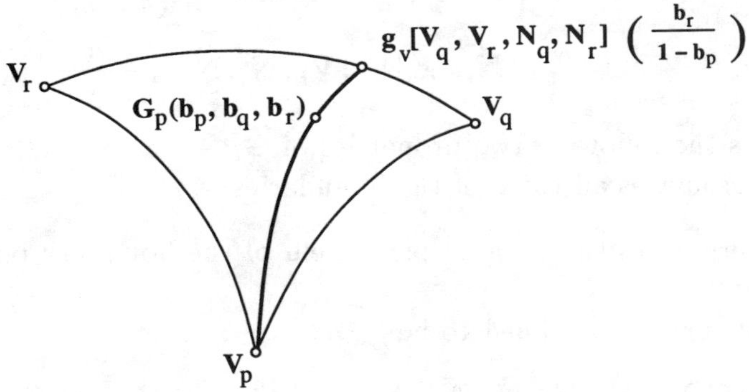



FIG. 8.5. *Side-vertex method.*

boundary curves together with a tangent plane field along each of these curves. The only requirements on each input curve are that it is $C^1$, and that it meet the other curves with a consistent tangent plane at each vertex. The tangent plane fields are specified using a normal vector field rather than a field of vectors in the tangent plane. That is, at each point along a boundary curve, the tangent plane is the plane perpendicular to the corresponding vector of the normal field. The normal fields are also required to be $C^1$, with the further restrictions that they must be nonzero everywhere and meet $C^0$ at the vertices. A surface patch is then constructed that matches this data, using a *side-vertex* method similar to that of the second scheme.

The second scheme is a side-vertex method that fits into the problem domain of this survey. The method proceeds by first constructing three boundary curves, one corresponding to each edge of the input triangle. Three patches are created, one for each boundary/opposite vertex pair. The interior of each patch is constructed by passing curves from points along the boundary (or "side") to the opposite vertex. Hence the name "side-vertex," as shown in Fig. 8.5. The three patches are then blended together to form the final patch.

All curves are constructed from two points and associated normals. We assume the existence of a curve construction operator $\mathbf{g}_v$ that takes two vertices with normals and constructs a curve:

$$\mathbf{g}_v[\mathbf{V}_0, \mathbf{V}_1, \vec{\mathbf{N}}_0, \vec{\mathbf{N}}_1](t),$$

such that $\mathbf{g}_v(0) = \mathbf{V}_0$, $\mathbf{g}_v(1) = \mathbf{V}_1$, $\mathbf{g}_v'(0) \cdot \vec{\mathbf{N}}_0 = 0$, and $\mathbf{g}_v'(1) \cdot \vec{\mathbf{N}}_1 = 0$. We will also assume the existence of a normal field constructor $\mathbf{g}_n$ that constructs a continuous normal field along the curve $\mathbf{g}_v$, where $\mathbf{g}_n$ is required to interpolate $\vec{\mathbf{N}}_0$ and $\vec{\mathbf{N}}_1$ at the end points.

The construction proceeds by building three patches, $\mathbf{G}_i$, $i \in \{p, q, r\}$

defined as:

$$\mathbf{G}_i(b_p, b_q, b_r) \;=\; \mathbf{g}_v\left[\mathbf{V}_i,\; \mathbf{g}_v[\mathbf{V}_j, \mathbf{V}_k, \vec{\mathbf{N}}_j, \vec{\mathbf{N}}_k]\left(\frac{b_k}{1 - b_i}\right),\right.$$

$$\left.\vec{\mathbf{N}}_i,\; \mathbf{g}_n[\mathbf{V}_j, \mathbf{V}_k, \vec{\mathbf{N}}_j, \vec{\mathbf{N}}_k]\left(\frac{b_k}{1 - b_i}\right)\right](1 - b_i).$$

Nielson notes the following two properties of $\mathbf{G}_i$:

1. $\mathbf{G}_i$ interpolates all three of the boundaries.

2. $\mathbf{G}_i$ interpolates the tangent plane field of the boundary opposite vertex $\mathbf{V}_i$.

The final surface is defined to be

$$\mathbf{G}[\mathbf{V}_p, \mathbf{V}_q, \mathbf{V}_r, \vec{\mathbf{N}}_p, \vec{\mathbf{N}}_q, \vec{\mathbf{N}}_r] = \beta_p \mathbf{G}_p + \beta_q \mathbf{G}_q + \beta_r \mathbf{G}_r,$$

where

(8.8)
$$\beta_i = \frac{b_j b_k}{b_p b_q + b_q b_r + b_r b_p}.$$

Nielson shows that the $\beta_i$ are such that the blending of any three surfaces having the two above properties yields a surface that interpolates all of the boundary curves and tangent fields. The theorem is reasonably general as it is true for a large class of $\mathbf{g}_v$ and $\mathbf{g}_n$. The operator $\mathbf{g}_v$ that Nielson presents constructs tangent vectors from the two normals and interpolates these two points and vectors with a cubic polynomial curve. The construction in Nielson's paper is not scale invariant, however, since the tangent vectors at the ends of the curve are normalized to unit vectors. This introduces loops in the curves if the data points are close together. The tangents should instead be scaled to be proportional in length to the distance between $\mathbf{V}_0$ and $\mathbf{V}_1$.

**Triangular Gregory Patches.**   Triangular Gregory patches are a variant of Gregory squares [14]. The key idea is that the twist term at the vertices is a blend of two twists, one for each boundary curve incident to the vertex. The scheme we present here is due to Longhi [22].

After constructing cubic boundary curves, this scheme uses the method of Chiyokura and Kimura to find a pair of cross-boundary control points for each edge (Fig. 8.6). In this figure, points $\mathbf{I}_{ij}$ and $\mathbf{I}_{ik}$ are the points constructed for the boundary associated with $b_i = 0$. When evaluating the patch at the domain point $(b_p, b_q, b_r)$, the two interior control points near each of the corner vertices are blended to form a single vertex, and thus, the six interior vertices are reduced to three control points. Points $\mathbf{I}_{ij}$ and $\mathbf{I}_{ik}$ are blended to produce the control point $\mathbf{I}_i$ using the following blend:

$$\mathbf{I}_i = \frac{b_k(1 - b_j)\mathbf{I}_{ik} + b_j(1 - b_k)\mathbf{I}_{ij}}{b_k(1 - b_j) + b_j(1 - b_k)}.$$

This yields a quartic Bézier patch, which is evaluated at $(b_p, b_q, b_r)$ to give a point on the surface.

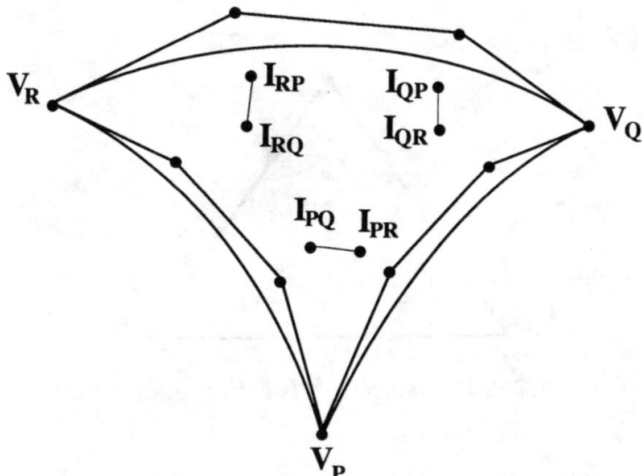


FIG. 8.6.   *Control points of a triangular Gregory patch.*

Triangular Gregory patches can also be thought of as a convex combination scheme. By putting the blending functions used to construct the $\mathbf{I}_i$s over a common denominator, the scheme can be rewritten as a convex combination of seven quartic Bézier patches. In this form, the blending functions are sixth degree rational polynomials, four degrees higher than the ones used by Nielson.

**Herron.**   In [19], Herron introduced a triangular surface fitting scheme in the following form:

$$\mathbf{F} = \mathcal{C} + b_p b_q b_r \mathcal{X},$$

where $\mathcal{C}$ interpolates the boundary curves and $\mathcal{X}$ is presented as a function that adjusts the cross-boundary tangents of $\mathcal{C}$ to meet specified tangent plane fields. Although it can be shown that $\mathbf{F}$ is a point-valued function, neither $\mathcal{C}$ nor $\mathcal{X}$ represent affine geometric entities (points, vectors, etc.).

We present here an alternative description of Herron's method that is more geometric in nature. We first observe that $\mathbf{F}$ can be rewritten as

$$(8.9) \qquad\qquad \mathbf{F} = \sum_{i=p,q,r} \beta_i \mathbf{F}_i,$$

where

$$\beta_i = \frac{b_j b_k}{b_i b_j + b_j b_k + b_k b_i},$$

and where each $\mathbf{F}_i$ is a quartic Bézier patch whose construction is given below. Note that these $\beta_i$ are identical to those used by Nielson (see (8.8)).

The input required by Herron's scheme is a triangle of points and the six boundary curve tangents at those points. Cubic Hermite interpolation is used on the boundary data to construct the boundary curves of $\mathbf{F}_i$. These curves have to be degree raised, as $\mathbf{F}_i$ is a quartic patch. This sets all the exterior control points for $\mathbf{F}_i$, leaving only the three interior control points, $\mathbf{C}_P$, $\mathbf{C}_Q$, and $\mathbf{C}_R$, to be determined (Fig. 8.7).

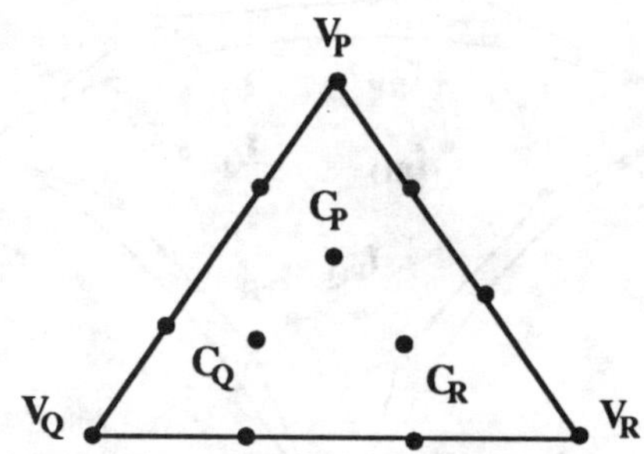

FIG. 8.7.  *Bézier patch for Herron's scheme.*

For patch $\mathbf{F}_i$, points $\mathbf{C}_j$ and $\mathbf{C}_k$ are constructed by using the triangular version of Chiyokura-Kimura. Equations 8.4 and 8.5 of §8.4.3 give formulas for these two points. $\mathbf{C}_i$, the final control point of $\mathbf{F}_i$, can be considered a free parameter. Herron's setting for this parameter is given in the Appendix.

Herron's scheme, then, can be thought of as a hybrid scheme using the cross-boundary tangent method of Chiyokura and Kimura for constructing triangular Bézier patches and the weighting functions of Nielson to produce the final patch. Note that we may use Nielson's proof above to show that $\mathbf{F}$ interpolates the tangent plane fields along all edges.[3]

A third way to view Herron's scheme is as a "three point" triangular Gregory patch. By rewriting (8.9) in Bernstein form, it is easily seen that Herron's method constructs a quartic Bézier patch, where each internal control point is a blend of three of the internal control points of the $\mathbf{F}_i$s. In Bézier form, then, the three point triangular Gregory patch blends together fewer patches than the two point Gregory patch (three patches instead of seven patches) and uses rational quadratic blending functions instead of the sextic rational polynomials of the two point scheme.

**8.5.3.  Other Schemes.**   There are other parametric surface fitting schemes that fit within the scope of this survey. Among these are one proposed by Farin [8], one proposed by Gregory and Charrot [15], and one proposed by Hagen and Pottmann [18].

Farin's scheme is a split domain scheme that is noteworthy primarily because it was the first parametric triangular surface fitting scheme. The construction, however, has several problems. One problem is that it is asymmetric in its treatment of the neighborhood of control points surrounding a vertex. This asymmetry is visible in the constructed surfaces, so we chose not to discuss Farin's scheme in detail here.

The method of Gregory and Charrot was originally intended by the authors to be used to fill triangular holes in an array of rectangular tensor product patches. Their scheme is a convex combination scheme that assumes cross-

---

[3]It is interesting to note that Herron's development of the method predated the publication of Chiyokura and Kimura [4] and Nielson [26].

boundary tangent fields have already been constructed along the boundary curves. Further, these tangent fields must admit a consistent mixed partial. Extending this scheme to fit into our problem domain would have been a fairly significant change. Although Gregory [16] later extended this method to allow for inconsistent mixed partials, we realized this too late to include it in this survey.

The discretized interpolant presented by Hagen and Pottmann [18] is another method that falls within our survey. This method extends Nielson's side-vertex method [26] and earlier work by Hagen [17] to second order geometric continuity. It is unfortunate that we learned of this method too late to include it in our survey.

## 8.6. Comparison

### 8.6.1. Tested Characteristics.
Our primary concern in this survey was with the visual appearance of the constructed surfaces. Other concerns, such as computational issues, were considered secondary, as we first wanted to find methods that produced nice shapes. Numerical stability issues are occasionally mentioned, as they can have a large impact on the shape of the resulting surface.

One problem with using visual appearance as our criterion is that it is a subjective measure of surface quality. In part, this stems from a lack of a general purpose "surface quality metric," that is, a commonly agreed upon definition of good shape. However, the problems with the shapes of surfaces we encountered were extreme, leaving little doubt as to the poor quality of the surfaces.

In many applications, the data points themselves are not distinguished points on the surface. Therefore, these points should not be visually distinguishable in the constructed surface. All schemes surveyed in this paper construct piecewise surfaces that have second order derivative discontinuities at the boundaries of the surfaces patches, implying that the boundaries of the patches (and, thus, the data points) will be distinguished to some extent. We feel that the visual impact of these discontinuities should be minimized.

In the past, many authors have used line drawing renditions to show the visual quality of their surfaces. We have found shaded images more useful in detecting various shape defects. For example, the line drawing in Fig. 8.8 is a plot of isoparametric lines of the Clough-Tocher interpolant to a function defined as the sum of three Gaussian functions. A shaded image of the same surface (Fig. 8.9) is far more informative. To see more subtle defects we found that Gaussian curvature plots of the interpolants were often helpful. (The Gaussian curvature at a point on a surface is the product of the minimum and maximum normal curvature of the surface at the point.)

All implemented schemes produced surfaces with shape defects that were readily apparent in the shaded images or Gaussian curvature plots. However, a scheme should not be considered "good" just because it passes these two visual

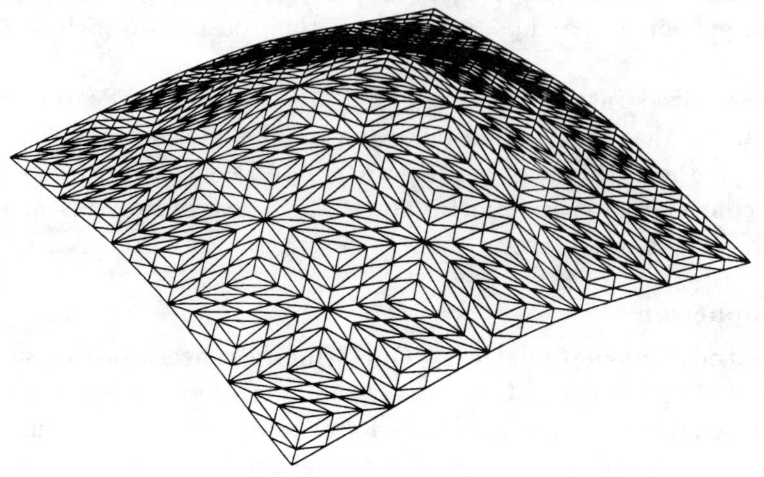

FIG. 8.8.   *Line drawing of Clough-Tocher surface.*

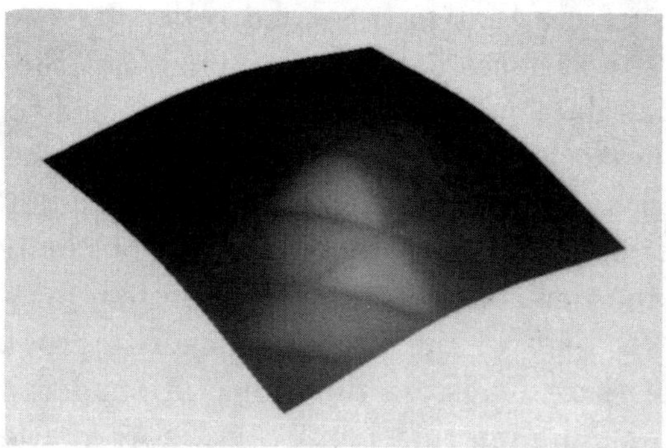

FIG. 8.9.   *Clough-Tocher interpolant constructed using estimated normals.*

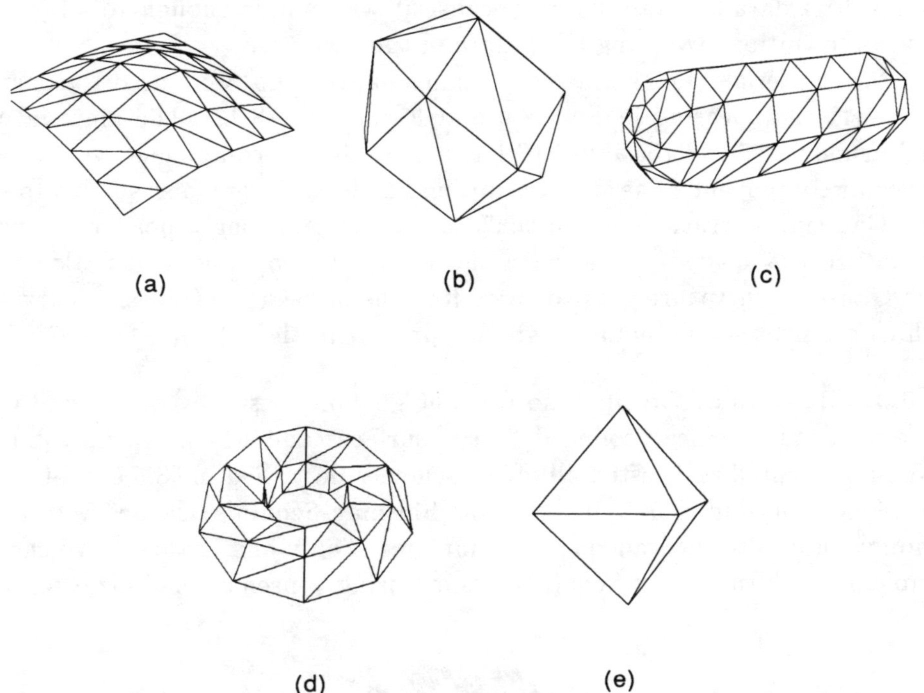

FIG. 8.10. *Some of the data sets used:* (a) *a Franke function,* (b) *sphere,* (c) *capsule,* (d) *torus, and* (e) *octahedron.*

tests. Until a good quantitative measure of shape is devised, a surface fitting scheme should also be tested by a variety of other methods, such as reflection lines and isophotes [30], to determine surface quality.

**8.6.2. Data Sets.**   A variety of data sets were used to test the surface fitting schemes. Several of these are shown in Figs. 8.10a–e. In these figures, the lines represent the edges of the triangulated data. The data points are located at the intersection of the lines.

One of the data sets is a sampling of one of the so-called Franke functions [13] (Fig. 8.10a). The underlying function is $z = \frac{1}{3}e^{-\frac{81}{4}[(x-.5)^2+(y-.5)^2]}$. The data sets of the sphere and the torus are samplings of those surfaces; the vertices and tangent planes of the octahedron data set have also been sampled from a sphere. The "capsule" data set is a sampling of a truncated cylinder with hemispherical caps.

The sphere data sets tended to be particularly "mild," as the entire surface has positive Gaussian curvature, that is, it has no flat spots or saddle points. The capsule data set was constructed to see if "ringing" would occur along the boundaries between the cylinder and the hemispheres. The torus is probably the most interesting data set, as it has regions of positive, negative, and zero Gaussian curvature. One problem with the dense data sets is that they are somewhat complex, making the resulting surfaces difficult to analyze. The octahedron data set was chosen because it was simple enough to allow us to develop intuition governing the failure of the schemes.

Two materials (i.e., surface reflectance parameters) were used to construct the surfaces appearing in Fig. 8.9 and Figs. 8.11– 8.17. The material used in Figs. 8.9, 8.11, 8.16, and 8.17 has a gray diffuse component with a white specular component. The surfaces in Figs. 8.12–8.15 were false shaded to show the Gaussian curvature of the surface. Areas of strongly positive Gaussian curvature are shaded white, with the intensity dropping to a dark gray as the Gaussian curvature goes to zero. Regions of negative Gaussian curvature, which occur at saddle points, were not present in these figures.

**8.6.3.  Results of Comparison.**   The goal of our survey was to find which interpolation schemes produced "nice" surfaces and which schemes did not. We implemented and tested all of the schemes described in §8.5 except for the three mentioned in §8.5.3. Jensen's and Shirman-Séquin's schemes were similar enough that it seemed adequate to implement only one of them. We chose to implement Shirman and Séquin's scheme using Jensen's generalization of the

FIG. 8.11.  *Shirman-Séquin interpolant constructed for octahedron data set.*

FIG.   8.12.    *Gaussian curvature plot of the Shirman-Séquin interpolant constructed for octahedron data set (Fig. 8.11).*

FIG. 8.13. *Gaussian curvature plot of the interpolant constructed for octahedron data set using de Boor-Höllig-Sabin method for computing boundary curves.*

FIG. 8.14. *Gaussian curvature plot of the Shirman-Séquin interpolant constructed for sphere data set.*

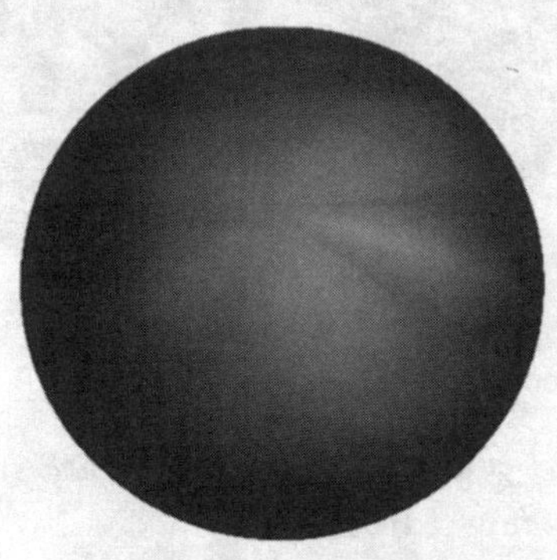

FIG. 8.15. *Gaussian curvature plot of the interpolant constructed for sphere data set using de Boor-Höllig-Sabin method for computing boundary curves.*

FIG. 8.16.  *Shirman-Séquin interpolant constructed for torus data set.*

FIG. 8.17.  *Interpolant constructed for torus data set using de Boor-Höllig-Sabin method for computing boundary curves.*

cross-boundary tangents. The software we developed to test these schemes is discussed elsewhere [24]. To our surprise, all of the schemes we tested performed rather poorly. Moreover, they all suffered from shape defects that are qualitatively similar.

**Scalar Data.**   Initially, we used sparse samplings of some of the Franke functions [13] for our data sets. Running two scalar data schemes on this input, we noticed several problems. First, as has been noted by many others (cf. [11], [12]), we found that the estimation of normals is a difficult problem, and second, these schemes fail to produce nice surfaces on data with high variation.

We decided not to address the problem of normal estimation, choosing to focus instead on performance of the methods once normals had been determined. Even when using normals sampled from a known surface, the shapes of the interpolants were still rather poor. As expected, most of the schemes seemed to perform better on data with less variation, that is, data that was nearly planar.

**Parametric Data.**   When we started working with parametric data schemes we looked at the interpolants for the data sets shown in Figs. 8.10b–e. The shaded images of the interpolants constructed by most schemes for the dense data sets usually have acceptable visual appearance, but Gaussian curvature plots indicate that there are subtle problems with them.

For example, inspection of Gaussian curvature plots for interpolants to the sphere data reveals that the patches are mostly flat, with a few areas of high curvature (Fig. 8.14). These effects occurred fairly uniformly for most schemes, with additional curvature discontinuities appearing along the interior boundaries produced by split domain schemes that are not present in the convex combination schemes. As a representative scheme, we show pictures of the surfaces constructed by Shirman and Séquin's scheme.

The one scheme that had additional kinds of difficulties to those problems mentioned above was Piper's scheme. Surfaces constructed by this scheme often exhibit displeasing undulations near the patch boundaries. This appears to be a result of numerical instabilities, occurring when the scalar functions of (8.2) are nearly constant (and, thus, when the $3 \times 5$ system of equations nearly reduces to a $2 \times 5$ system). We decided not to address these numerical stability issues, focusing instead on the other schemes.

Shaded images of the interpolants to the torus data revealed problems more serious than the ones mentioned above (Fig. 8.16). Images of Gaussian curvature indicated that there were large variations of curvature, even within a single patch. Discontinuous jumps from positive to negative curvature were also observed along patch boundaries.

Images of Gaussian curvature for interpolants to the simple data set of the octahedron (Fig. 8.11) clearly show that curvature is concentrated near the

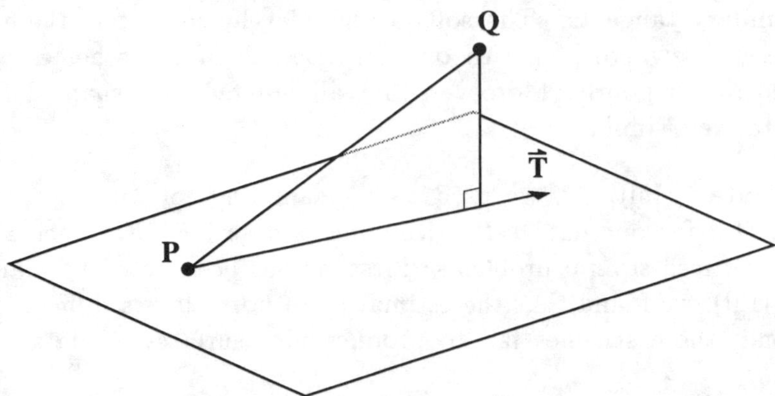

FIG. 8.18.    *The construction of the tangent at* **P** *for the curve from* **P** *to* **Q**. $|\vec{\mathbf{T}}| = |\mathbf{P} - \mathbf{Q}|$.

vertices and boundary curves (Fig. 8.12). The center of the patch (or patches) created for a face tend to be relatively flat by comparison. These problems were similar for all the schemes, which was somewhat unexpected, considering that the split domain schemes construct surfaces in a very different fashion from the convex combination schemes.

Some of these problems could be alleviated by manually adjusting the scalar shape parameter of Jensen's scheme. For example, the curvature of the surfaces constructed for the octahedron could be spread over the patches somewhat more uniformly, but there are still flat spots on these surfaces. The improvements possible for the torus data set are less significant. It is also unclear how to set automatically the value of this parameter.

**8.6.4.  Boundary Curves.**  The construction of boundary curves was a common element in all our implementations.  Following constructions given by Piper and Shirman-Séquin, we originally used the following method for constructing a cubic boundary curve for an edge of the data set: The end points of the curve are set to interpolate the end points of the edge. Next, tangent directions are computed for each end point by perpendicularly projecting the vector along the edge into the tangent plane at each end point; these vectors are then scaled to be of length equal to the length of the edge (see Fig. 8.18). The boundary curve is then set to be the cubic polynomial curve matching this data.

As published, the surveyed schemes used a variety of methods for constructing boundary curves. However, the differences between these methods are minor. Shirman and Séquin use the method mentioned above. Herron and Jensen essentially assume that the boundary curves have already been constructed. Nielson constructs cubic boundary curves whose tangents agree in direction with the tangents in the above construction, but leaves the length as a free parameter. Piper's scheme is unique among these schemes in that

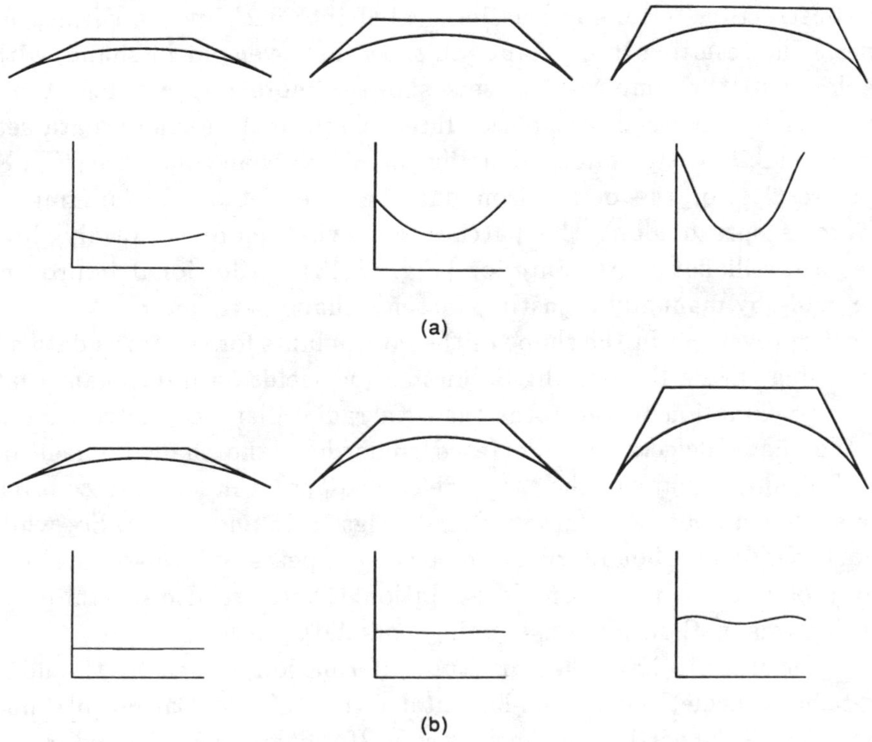

FIG. 8.19. *Bézier curves and their curvature plots.* (a) *Curves constructed using projection method.* (b) *Curve constructed by de Boor-Höllig-Sabin method.*

it constructs quartic boundary curves; the end point tangents have the same direction as those in the method presented above, but are shorter in length.

Curvature plots of boundary curves constructed as above showed that as the tangent vectors become more perpendicular to the vector along the edge the curvature along the curve is concentrated near the end points (Fig. 8.19a), leaving a relatively flat region in the middle of the curve. In all the surface schemes we implemented, this flatness was then propagated inward in the patch construction, resulting in relatively large flat areas in the middle of the patches. This suggested that it might be possible to improve the curvature distribution of the patches by improving the curvature distribution of the boundary curves.

In a paper by de Boor, Höllig, and Sabin [5], a method is given for constructing planar cubic curves that interpolate to positions, tangent lines, and curvatures at two end points. More importantly for our purposes, these curves were observed to have relatively uniform curvature distributions. In Fig. 8.19b, for example, the curvature values are taken from the circle passing through the data points with the given tangents. If the curvature values at the end points had not been equal, the resulting curves would have been approximations to an ellipse.

To determine if using boundary curves with a more uniform distribution of curvature would yield surfaces with better shape, the de Boor-Höllig-Sabin curve construction technique was integrated into each of the surface fitting schemes. The resulting interpolants all show improvement in shape, with the interpolants to the simpler data sets showing more improvement than the interpolants to more complex ones. Interpolants to the sphere data set, for example, exhibit nearly uniform distribution of Gaussian curvature (Fig. 8.15).

The results for the octahedron data set are not as encouraging. The curvature is spread along the patches somewhat more uniformly, but the patches are still flat in the interior (Fig. 8.13). Additional improvements can be made by manually adjusting Jensen's shape parameter.

The improvement in the shape of the interpolants for the torus data set are minor. Using the de Boor-Höllig-Sabin method yields an interpolant that is a better approximation to the torus (using a radial distance metric); however, while the shape defects are alleviated somewhat, the shaded image of the surface still shows many of the shape defects apparent in the surface produced by the standard boundary curve method (Figs. 8.16 and 8.17). So, while the de Boor-Höllig-Sabin boundary curve method appears to improve the shape of the interpolants, it is not a complete solution. There are also several problems with using this method for constructing boundary curves.

First, in order to have the curvature information needed by the de Boor-Höllig-Sabin scheme, second fundamental forms (cf. do Carmo [6]) must be associated with the vertices of the data sets. If the data are sampled from a $C^2$ function, then second fundamental forms can be calculated directly. If only the data points are available, then second fundamental forms must be estimated. It is not clear at this time how difficult it is to make these estimates.

A more serious problem is that there are from zero to three cubic curves which match the curvature data [5]. For our purposes, if there are zero curves that interpolate the data, then the boundary curves must be computed by some other method. If there are multiple solutions, then a choice must be made between these solutions.

The case of multiple solutions revealed another interesting fact. In the cases we saw, all of the solution curves given by the de Boor-Höllig-Sabin method have essentially the same shape. These solutions differ primarily in their parameterizations. For the sphere data, for example, there are three distinct cubic curves that match the end point data. The solution that is most nearly uniform in parameterization gives the surface shown in Fig. 8.15. However, when one of the less uniform parameterizations is used, large lumps appear in the surface. Thus, both the shape and the parameterization of the boundary curves are important.

Another issue concerning the use of the de Boor-Höllig-Sabin method for constructing boundary curves is that it mandates the use of planar curves. Thus, a plane must be chosen in which to place each boundary curve. The choice of this plane can be thought of as a free parameter. The plane we used is the one containing the edge, and whose normal vector is the cross product of the vector along the edge and the average of the normals at the end points.

It is unclear how restricting the boundary curves to lie in a plane affects the shape of the surfaces.

## 8.7. Summary

At the beginning of our study we did not expect to find a method that would work well for arbitrarily placed data. It soon became clear, however, that we had drastically underestimated the difficulty of the problem. As expected, the schemes produced poor interpolants to extremely sparse data sets. More surprising was how hard it was to produce good interpolants for any but the most benign data sets. In particular, we observed the following:

- Although different schemes constructed surfaces in different fashions, all surfaces displayed similar shape defects.

- The primary cause of the shape defects appears to be in the construction of boundary curves. Flatness on the boundary curves is propagated inward, resulting in flat spots on the surface.

The shape and parameterization of the boundary curves greatly influences the shape of the patches. To construct surfaces with better shape, a curve construction method must be found that spreads the curvature uniformly along the boundary curves. One way to construct such curves might be to relax the locality condition on the schemes and do some form of global optimization on the boundary curves, perhaps something similar to Nielson's minimum norm networks [25], [27]. Alternatively, a local method based on curvature interpolation and de Boor-Höllig-Sabin's method seems promising, at least for approximating known surfaces. A disadvantage of such an approach is that good methods of estimating second fundamental forms would need to be developed. It is also unclear what ramifications the restrictions to planar boundary curves might have.

Improving the shape of boundary curves, however, is not a complete solution to the construction of surfaces free of unnecessary shape defects. Many schemes provide additional shape parameters that can be adjusted to improve the appearance of the interpolant. Such shape parameters may be useful in interactive design applications, but it is important to develop methods for setting good default values. In the approximation of known surfaces, it is also important to understand how differential geometric properties of the known surface can be used to set the free parameters.

## 8.8. Recommendations

Since all of the schemes produce surfaces with similar shapes, we recommend using the Shirman-Séquin scheme, but only because it constructs polynomial patches rather than the rational polynomial patches built by most of the other schemes. The use of polynomial patches simplifies the calculation of derivatives, curvature, etc., of the interpolant. Although the domain split introduces extra artifacts in the surfaces, such as the creation of long, thin triangles, the major shape defects are common to all schemes.

Triangular Gregory patches are often suggested as the scheme of choice. However, it is hard to determine differential information since the patches are rational polynomials of fairly high degree. Further, we found that this scheme is extremely sensitive to the settings of the free parameters. While this might be desirable in some situations, it appears to be difficult to control these shape parameters and to determine reasonable default values for them.

Estimating second fundamental forms at the data points appears to have two benefits. First, the second fundamental form can be used to solve the vertex consistency problem. It can also be used to produce boundary curves with more uniform curvature, resulting in surfaces with better shape. When approximating known surfaces, the surface can be sampled for position, tangent, and second fundamental form (assuming it is $C^2$). Using all of this information in the construction of the approximating surface, a higher order of convergence should be achievable.

## Acknowledgments

This work was supported in part by the National Science Foundation under grant numbers DMC-8802949, CCR-8957323, CCR-8612543, and IRI-8801932. Support from the Digital Equipment Corporation, Xerox, and IBM is also gratefully acknowledged.

## Appendix

This appendix gives Herron's setting of the control point $\mathbf{C}_i$ (Fig. 8.7). We will denote the tangent from point $\mathbf{V}_i$ to $\mathbf{V}_j$ as $\vec{\mathbf{T}}_{ij}$. In the construction of patch $\mathbf{F}_i$, Herron implicitly sets $\mathbf{C}_i$ to be:

$$\mathbf{C}_i \;=\; \tfrac{1}{12}\Big\{ -6\mathbf{V}_j - 2\vec{\mathbf{T}}_{ji} + \vec{\mathbf{T}}_{ik} - \vec{\mathbf{T}}_{ij} + \tfrac{9}{4}\Big[\vec{\mathbf{R}}(\tfrac{1}{3}) - \mathcal{Q}(\tfrac{1}{3})\Big] $$
$$+ \tfrac{9}{4}\Big[\vec{\mathbf{R}}(\tfrac{2}{3}) - \mathcal{Q}(\tfrac{2}{3})\Big]\Big\} + \mathbf{E}_{i2}^4,$$

where

$$\mathbf{E}_{i2}^4 = \frac{1}{2}\mathbf{V}_j + \frac{1}{6}\vec{\mathbf{T}}_{ji} + \frac{1}{2}\mathbf{V}_i + \frac{1}{6}\vec{\mathbf{T}}_{ij},$$

and $\vec{\mathbf{R}}(t) = k(t) \cdot \vec{\mathbf{C}}(t)$ (from Section 8.4.3) and

$$\mathcal{Q}(t) \;=\; 6t(1-t)p_j(t) \cdot \mathbf{V}_j + 6t(1-t)p_k(t) \cdot \mathbf{V}_k$$
$$+ [(1-t)^2 p_k(t) + 2t(1-t)p_j(t)] \cdot \vec{\mathbf{T}}_{jk}$$
$$+ [t^2 p_j(t) + 2t(1-t)p_k(t)] \cdot \vec{\mathbf{T}}_{kj}$$
$$+ (1-t)^2 \vec{\mathbf{T}}_{ji} + t^2 \vec{\mathbf{T}}_{ki}.$$

Here $p_j(t)$ and $p_k(t)$ are the scalar functions

$$p_j(t) = t(r_{ji} + r_{ki} + 1) - r_{ji} - 1$$

and

$$p_k(t) = (1-t)(r_{ji} + r_{ki} + 1) - r_{ki} - 1,$$

where $r_{ji}$ and $r_{ki}$ are:

$$r_{ji} = \frac{\vec{\mathbf{T}}_{jk} \cdot \vec{\mathbf{T}}_{ji}}{\vec{\mathbf{T}}_{jk} \cdot \vec{\mathbf{T}}_{jk}} \quad \text{and} \quad r_{ki} = \frac{\vec{\mathbf{T}}_{kj} \cdot \vec{\mathbf{T}}_{ki}}{\vec{\mathbf{T}}_{kj} \cdot \vec{\mathbf{T}}_{kj}}.$$

## References


[1] R. Barnhill, *Computer aided surface representation and design*, in Surfaces in Computer Aided Geometric Design, R. Barnhill and W. Boehm, eds., North-Holland, Amsterdam, 1983, pp. 1–24.

[2] W. Boehm, *Visual continuity*, Comput. Aided Des., 20 (1988), pp. 307–311.

[3] H. Chiyokura, *Localized surface interpolation method for irregular meshes*, in Advanced Computer Graphics, Tosiyasu L. Kunii, ed., Springer-Verlag, New York, 1986, pp. 3–19.

[4] H. Chiyokura and F. Kimura, *Design of solids with free-form surfaces*, Computer Graphics, 17 (1983), pp. 289–298.

[5] C. de Boor, K. Höllig, and M. Sabin, *High accuracy geometric Hermite interpolation*, Comput. Aided Geom. Des., 4 (1987), pp. 269–278.

[6] M. do Carmo, *Differential Geometry of Curves and Surfaces*, Prentice-Hall, Englewood Cliffs, NJ, 1976.

[7] G. Farin, *A construction for visual $C^1$ continuity of polynomial surface patches*, Computer Graphics and Image Processing, 20 (1982), pp. 272–282.

[8] ———, *Smooth interpolation to scattered $3D$ data*, in Surfaces in Computer Aided Geometric Design, R. Barnhill and W. Boehm, eds., North-Holland, Amsterdam, 1983, pp. 43–63.

[9] ———, *A modified Clough-Tocher interpolant*, Comput. Aided Geom. Des., 2 (1985), pp. 19–27.

[10] ———, *Curves and Surfaces for Computer Aided Geometric Design*, Academic Press, New York, 1988.

[11] R. Franke, *Scattered data interpolation: Tests of some methods*, Math. Comp., 38 (1982), pp. 181–200.

[12] R. Franke and G. Nielson, *Scattered data interpolation: A tutorial and survey*, in Geometric Modeling: Methods and Applications, H. Hagen and D. Roller, eds., Springer-Verlag, New York, 1991, pp. 131–160.

[13] T. Grandine, *An iterative method for computing multivariate $C^1$ piecewise polynomial interpolants*, Comput. Aided Geom. Des., 4 (1987), pp. 307–319.

[14] J. Gregory, *Smooth interpolation without twist constraints*, in Computer Aided Geometric Design, R. Barnhill and R. Riesenfeld, eds., Academic Press, New York, 1974, pp. 71–87.

[15] J. Gregory and P. Charrot, *A $C^1$ triangular interpolation patch for computer-aided geometric design*, Computer Graphics and Image Processing, 13 (1980), pp. 80–87.

[16] J. Gregory, *N-sided surface patches*, in The Mathematics of Surfaces, J. Gregory, ed., Clarendon Press, Oxford, 1986, pp. 217–232.

[17] H. Hagen, *Geometric surface patches without twist constraints*, Comput. Aided Geom. Des., 3 (1986), pp. 179-184.

[18] H. Hagen and H. Pottmann, *Curvature continuous triangular interpolants*, in Mathematical Methods in Computer Aided Geometric Design, T. Lyche and L. L. Schumaker, eds., Academic Press, New York, 1989, pp. 373–384.

[19] G. Herron, *Smooth closed surfaces with discrete triangular interpolants*, Comput. Aided Geom. Des., 2 (1985), pp. 297–306.

[20] ———, *Techniques for visual continuity*, in Geometric Modeling: Algorithms and New Trends, G. Farin, ed., Society for Industrial and Applied Mathematics, Philadelphia, 1987, pp. 163–174.

[21] T. Jensen, *Assembling triangular and rectangular patches and multivariate splines*, in Geometric Modeling: Algorithms and New Trends, G. Farin, ed., Society for Industrial and Applied Mathematics, Philadelphia, 1987, pp. 203–220.

[22] L. Longhi, *Interpolating patches between cubic boundaries*, Tech. Report T.R. UCB/CSD 87/313, University of California, Berkeley, Berkeley, CA 94720, October 1986.

[23] C. Loop, *A $G^1$ triangular spline surface of arbitrary topology*, in preparation.

[24] M. Lounsbery, C. Loop, S. Mann, D. Meyers, J. Painter, T. DeRose, and K. Sloan, *A testbed for the comparison of parametric surface methods*, in Curves and Surfaces in Computer Vision and Graphics, L. A. Ferrari and R. J. P. de Figueiredo, eds., SPIE Proceedings 1251, February, 1990, pp. 94–105.

[25] G. M. Nielson, *Minimum norm interpolation in triangles*, SIAM J. Numer. Anal., 17 (1980), pp. 44–62.

[26] ———, *A transfinite, visually continuous, triangular interpolant*, in Geometric Modeling: Algorithms and New Trends, G. Farin, ed., Society for Industrial and Applied Mathematics, Philadelphia, 1987, pp. 235–246.

[27] ———, *Interactive surface design using triangular network splines*, in International Conference on Engineering Graphics and Descriptive Geometry Proceedings, vol. 2, 1988, pp. 70–77.

[28] J. Peters, *Smooth interpolation of a mesh of curves*, Constr. Approx., 7 (1991), pp. 221–247.

[29] B. Piper, *Visually smooth interpolation with triangular Bézier patches*, in Geometric Modeling: Algorithms and New Trends, G. Farin, ed., Society for Industrial and Applied Mathematics, Philadelphia, 1987, pp. 221–233.

[30] T. Poeschl, *Detecting surface irregularities using isophotes*, Comput. Aided Geom. Des., 1 (1984), pp. 163–168.

[31] L. A. Shirman and C. H. Séquin, *Local surface interpolation with Bézier patches*, Comput. Aided Geom. Des., 4 (1987), pp. 279–295.

[32] J. J. van Wijk, *Bicubic patches for approximating non-rectangular control-point meshes*, Comput. Aided Geom. Des., 3 (1986), pp. 1–13.

# Free-Form Surfaces from Partial Differential Equations

M. I. G. Bloor and M. J. Wilson

## 9.1. Introduction

We have described previously [1] a method for surface generation whereby a smooth surface is generated as a solution to a boundary-value problem by means of an elliptic partial differential equation (PDE). The model equation with which we illustrated the principles of the method was based upon the biharmonic equation $\Delta^2 \phi = 0$. The method was introduced in the field of blend generation where it was demonstrated that a boundary-value approach was particularly suited to the problem of having to generate a smooth bridging surface between two "primary" surfaces. We wish in this paper to discuss how the method may be applied to the problem of free-form surface design.

The defining characteristic of a free-form surface is its property of having a shape that can be influenced by a designer. Such surfaces occur in computer aided geometric design (CAGD) in a variety of contexts ranging from the representation of existing objects through to shape design. Conventionally, free-form surfaces are usually bivariate polynomial functions of two surface parameters whose shape can be altered by manipulating an associated mesh of control points to which, in some sense, the surface is an approximation. The type of polynomial functions used gives the various "classical" forms of curve and surface representations e.g., Coons patches [2], Bézier curves and surfaces [2], [3], B-splines [2]–[4], rational B-splines [5], [6]; and free-form surface design using such curve-surface representations has been discussed by a number of authors [7], [8]. In contrast to this, PDE surfaces, being generated as solutions to boundary-value problems, can be manipulated by changing the boundary conditions imposed at the edges of the patch (in this respect they are similar to Bézier patches). Thus, it can readily be appreciated that from the conceptual point of view, the method is a departure from accepted practice in this area. What is new about the method is its philosophy regarding surface manipulation and the selection of surfaces with desirable properties, which is basically that of a boundary-value approach. However, from the purely practical point of view, the surfaces produced by the PDE method can be expressed, or rather approximated, in terms of conventional tensor product surfaces. In particular, it has been shown how to express a PDE surface in terms of B-splines [9]–[11].

There are certain similarities between the PDE method and other design methods, in particular the "energy-based" method described by Celniker and Gossard [12]. The basis of the technique is to regard a free-form surface as, in some sense, behaving as an elastic membrane that one can manipulate by applying an appropriate distribution of "loads." The behavior of this idealized membrane is governed by a PDE, hence the similarity with the present method. There are also similarities with the variational approach to surface generation (see, for example, Hagen and Schulze [13], Nowacki, Dingyuan, and Xinmin [14]), due to the fact that the elliptic PDEs with which we are concerned have an equivalent variational formulation. Note also the work of Williams [15], which links the variational approach to a structural analogy.

In spite of the elements which the PDE method shares with other approaches, what distinguishes it at the present time are certain technical points and, most importantly, the decoupling of the physical coordinates plus its emphasis upon the role of boundary conditions in the design of a surface. In this paper we wish to illustrate this by means of examples showing something of what may be achieved by the method, but, before we do so, we will outline its mathematical basis.

## 9.2. The Mathematical Basis of the PDE Method

In previous papers we have considered solutions to elliptic PDEs [16]. The general linear elliptic PDE may be written in the form

$$(9.1) \qquad \sum_{n=0}^{2r} \alpha_n(u,v) \frac{\partial^n}{\partial u^l \partial v^m} \underline{X}(u,v) = \underline{F}(u,v),$$

which represents a PDE of order $2r$, where $1, m, n \geq 0$, $1 + m = n$. Note that solutions to this type of equation are smooth in the sense of being infinitely differentiable and also, in the present context, give rise to surfaces that may be described as "fair." The term on the right-hand side of the equation we shall refer to as a forcing term (cf. Celniker and Gossard's loading function [12]) and may be used to produce local changes in surface shape [17]. However, in this paper, we shall only consider the case where $\underline{F}(u,v) = \underline{0}$ in order to illustrate what control over a surface's shape may be achieved with the boundary conditions.

### 9.2.1. "Harmonic" PDE Surfaces.

The simplest elliptic PDE is one based on Laplace's equation, thus

$$(9.2) \qquad \left( \frac{\partial^2}{\partial u^2} + a^2 \frac{\partial^2}{\partial v^2} \right) \underline{X}(u,v) = 0.$$

The solution to this equation $\underline{X}(u,v) = (x(u,v), y(u,v), z(u,v))$ is a vector-valued function of the two independent variables $u$ and $v$, which serve to define a coordinate system within the surface. The components of $\underline{X}$ are the Euclidean coordinate functions of the surface. The function $\underline{X}(u,v)$ is defined over some region $\Omega$ of the $(u,v)$ parameter plane and satisfies not only the PDE but also certain boundary conditions, on $\underline{X}$ or its normal derivative $\partial \underline{X}/\partial \mathbf{n}$ (in the $(u,v)$ parameter plane), for example. We may think of $\underline{X}$ as a (harmonic) mapping from a region of $(u,v)$ parameter space to some surface in $E^3$. Note that the boundary conditions on $\underline{X}$ basically determine how the space curves bounding the surface in $E^3$ are parameterized in terms of $u$ and $v$, in that, around the edges of the $(u,v)$ region $\underline{X}$, for instance, is specified as a function of $u$ and $v$.

An example of such a surface is shown in Fig. 9.1 where the boundary conditions are such that the surface begins and ends on two circular space curves. In order to obtain this surface, (9.2) has been solved independently three times, once for each component of $\underline{X}$. The boundary conditions are of the form

$$
(9.3) \quad
\begin{aligned}
x(0,v) &= r_1 \cos(v), & y(0,v) &= r_1 \sin(v), & z(0,v) &= h, \\
x(1,v) &= r_2 \cos(v), & y(1,v) &= r_2 \sin(v), & z(1,v) &= 0,
\end{aligned}
$$

where $r_1$, $r_2$, and $h$ are constants, and the PDE has been solved subject to periodic boundary conditions in the $v$ direction.

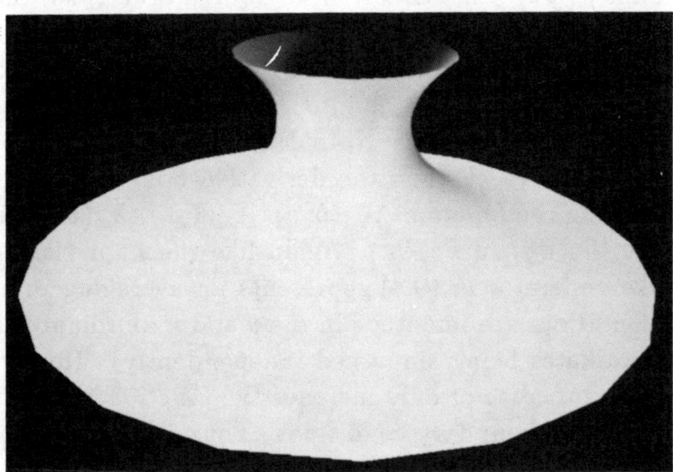

FIG. 9.1. *Laplacian PDE surface between two circular boundary curves.*

As for the shape of the surface, the Laplacian is an averaging operator and its action is to produce a smooth transition between the boundary conditions. Note that the surface is not unlike the "minimal-surface" produced by a soap-film stretched between the two circles. This is not altogether surprising since minimal surfaces actually satisfy (9.2) in the case of $a = 1$ [18]. However, although minimal surfaces often have aesthetically pleasing shapes, they are of rather limited use in surface design (however, see Rockwood [19] for a discussion of the use of such surfaces in shape design). For one thing, there is the nontrivial task of ensuring that the surface is parameterized in terms of a set of "isothermal" coordinates $u$ and $v$, while for another, minimal surfaces offer the designer limited scope for surface manipulation. This second objection is equally true in general for solutions to (9.2), and in fact for all such second-order PDEs (with zero right-hand side), since the boundary conditions needed to find a solution essentially only allow us to specify the shapes of the curves bounding the surface or else the direction of the tangent planes, but not both simultaneously at a point. This gives some control over the shape of the patch interior, especially if we vary the parameter $a$ or the parameterization of the boundary curves, but there is not the scope for shape control that freedom to specify both function and derivative boundary conditions would give us. It is for this reason that we shall concentrate our discussion upon higher-order PDEs, in particular fourth-order equations.

**9.2.2. Biharmonic PDE Surfaces.** In previous work we have considered solutions to the following elliptic PDE based upon the biharmonic equation

$$(9.4) \qquad \left( \frac{\partial^2}{\partial u^2} + a^2 \frac{\partial^2}{\partial v^2} \right)^2 \underline{X}(u,v) = 0.$$

Like (9.2), this equation is solved over some region $\Omega$ of the $(u,v)$ parameter plane, but unlike (9.2) this equation is fourth-order, so that boundary conditions on both $\underline{X}$ and its normal derivative $\partial \underline{X}/\partial \mathbf{n}$ can be specified when finding a solution. As before, the boundary conditions on $\underline{X}$ and $\partial \underline{X}/\partial \mathbf{n}$ are imposed by specifying these two quantities as functions of $u$ and $v$ around the boundary $\partial \Omega$ of the patch. Note that if the patch happened to be the unit square $[0,1] \times [0,1]$, then the derivative boundary conditions would amount to specifying the functions $\underline{X}_u(0,v)$, $\underline{X}_u(1,v)$, $\underline{X}_v(u,0)$, and $\underline{X}_v(u,1)$, where $\underline{X}_u = (\partial x/\partial u, \partial y/\partial u, \partial z/\partial u)$. Again like the Laplacian operator, this "biharmonic-like" operator in (9.4) represents an averaging process in which the boundary conditions are smoothed in the $u$ and $v$ coordinate directions (the $x$, $y$, and $z$ coordinates being smoothed independently). However, note that now the averaging process not only includes the function boundary conditions but also the derivative boundary conditions. Thus, we have extra control over the surface's shape from the derivative boundary conditions which determine the "velocity" of the $u$ and/or $v$ isoparametric lines as they leave the boundary curves.

The purely formal way in which the derivative boundary conditions influence the solution can be seen from (9.5), which expresses the solution for one of the components of $\underline{X}$ in terms of a boundary integral formulation [20]. If we denote the dependent variable by $x_i$, where $x_i$ may be either $x$, $y$, or $z$ in the present context, then the value of $x_i$ at any point $\mathbf{p}$ in $\Omega$ is given by the following integral evaluated around $\partial \Omega$,

$$(9.5) \qquad x_i(\mathbf{p}) = \frac{1}{2\pi} \int_{\partial \Omega} (x_i)_c \frac{\partial}{\partial n}(\Delta G) - \Delta G \left( \frac{\partial x_i}{\partial n} \right)_c ds,$$

where $\Delta$ is the "modified" Laplacian operator $(\partial^2/\partial u^2 + a^2\partial^2/\partial v^2)$; the subscript $c$ denotes values on the boundary $\partial\Omega$ and $\frac{\partial}{\partial n}$ denotes the partial derivative in the direction of the outward normal to the contour $\partial\Omega$. The Green's function, $G = G(\mathbf{p}, \mathbf{q})$, for the biharmonic equation at a fixed point $\mathbf{p}$ is a function of $\mathbf{q}$ in $\Omega$ and must satisfy

(i)     $\nabla^4 G = 0$ in $\Omega$,

(ii)    $G = 0$ and $\dfrac{\partial G}{\partial n} = 0$ on $\partial\Omega$,

(iii)   $G(\mathbf{p}, \mathbf{q}) \to r^2 \ln r$ as $r \to 0$, where $r = |\mathbf{p} - \mathbf{q}|$,

(iv)   $G(\mathbf{p}, \mathbf{q})$ has no other singularities in $\Omega$ or on $\partial\Omega$.

The integral on the right-hand side of (9.5) represents the smoothing of the function and its normal derivative specified on the boundary. At any point $\mathbf{p}$ in $\Omega$, $x_i$ is a weighted average of its value and the value of its normal derivative around the bounding contour.


**9.2.2.1. The Effect of the Derivative Boundary Conditions.** Inspection of the Green's function is not particularly instructive when it comes to an intuitive understanding of the role played by the derivative boundary conditions. It is easier to remember that these conditions essentially determine the "speed" ($|\underline{X}_u|$ or $|\underline{X}_v|$) and direction (in $E^3$) of the isoparametric lines as they leave the curves bounding the patch, and consider a few practical examples. In this section we will illustrate the effect of the derivative boundary conditions on the shape of simple four-sided surface patch.

Consider the solution of (9.4) which satisfies the following boundary conditions around the edges of the unit square in $(u, v)$ space.

$$(9.6) \qquad \begin{aligned} \underline{X}(u,0) &= (u,0,0), \quad \underline{X}(u,1) = (u,1,0), \\ \underline{X}(0,v) &= (0,v,0), \quad \underline{X}(1,v) = (1,v,0) \end{aligned}$$

$$(9.7) \qquad \begin{aligned} \underline{X}_v(u,0) &= (0,1,0), \quad \underline{X}_v(u,1) = (0,1,0), \\ \underline{X}_u(0,v) &= (1,0,0), \quad \underline{X}_u(1,v) = (1,0,0) \end{aligned}$$

The solution is nothing more complicated than a unit square in $E^3$ and is shown in Fig. 9.2. If we now keep the function boundary conditions (9.6) the same but change the derivative boundary conditions (9.7), we can alter the shape of the patch interior while keeping its boundary curves the same. Consider the surfaces shown in Figs. 9.3–9.6 for which the derivative boundary conditions (9.7) have been replaced by the conditions in (9.8)–(9.11), respectively. These solutions have all been generated numerically, by discretizing (9.4) using finite differences and then solving the resulting set of algebraic equations using SOR.

$$(9.8) \qquad \begin{aligned} \underline{X}_v(u,0) &= (0,1,4\sin(\pi u)), \quad \underline{X}_v(u,1) = (0,1,-4\sin(\pi u)), \\ \underline{X}_u(0,v) &= (1,0,4\sin(\pi v)), \quad \underline{X}_u(1,v) = (1,0,-4\sin(\pi v)), \end{aligned}$$

$$(9.9) \qquad \begin{aligned} \underline{X}_v(u,0) &= (0,1,4\sin(\pi u)), \quad \underline{X}_v(u,1) = (0,1,4\sin(\pi u)), \\ \underline{X}_u(0,v) &= (1,0,4\sin(\pi v)), \quad \underline{X}_u(1,v) = (1,0,4\sin(\pi v)), \end{aligned}$$

$$(9.10) \quad \begin{aligned} \underline{X}_v(u,0) &= (0,1,4\sin(2\pi u)), & \underline{X}_v(u,1) &= (0,1,-4\sin(2\pi u)), \\ \underline{X}_u(0,v) &= (1,0,4\sin(2\pi v)), & \underline{X}_u(1,v) &= (1,0,-4\sin(2\pi v)), \end{aligned}$$

$$(9.11) \quad \begin{aligned} \underline{X}_v(u,0) &= (0,1,4\sin(\pi u)), & \underline{X}_v(u,1) &= (0,1,-4\sin(\pi u)), \\ \underline{X}_u(0,v) &= (1,0,0), & \underline{X}_u(1,v) &= (1,0,0). \end{aligned}$$

FIG. 9.2.  *The effect of the derivative boundary conditions: Biharmonic PDE surface satisfying boundary conditions (9.6) and (9.7).*

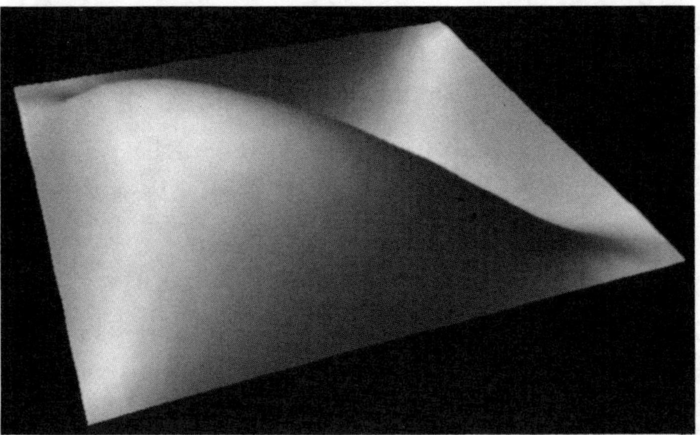

FIG. 9.3.  *The effect of the derivative boundary conditions: Biharmonic PDE surface satisfying boundary conditions (9.6) and (9.8).*

Notice how the boundary conditions on the vectors $\underline{X}_u$ and $\underline{X}_v$ control the shape of the patch interior.  In this simple example, the variation in these boundary coordinate vectors on the boundary has been achieved by a modification of their $z$-components.  In general we are free to alter all three components as we please in order to achieve the desired shape.  The next section contains more practical examples which illustrate the control we can

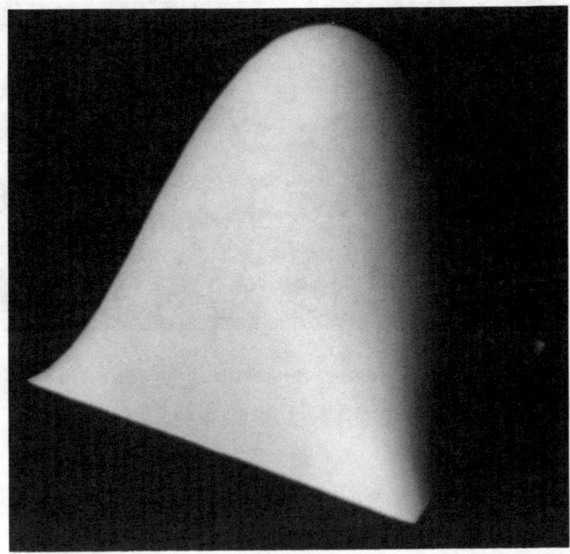

FIG. 9.4.   *The effect of the derivative boundary conditions: Biharmonic PDE surface satisfying boundary conditions (9.6) and (9.9).*

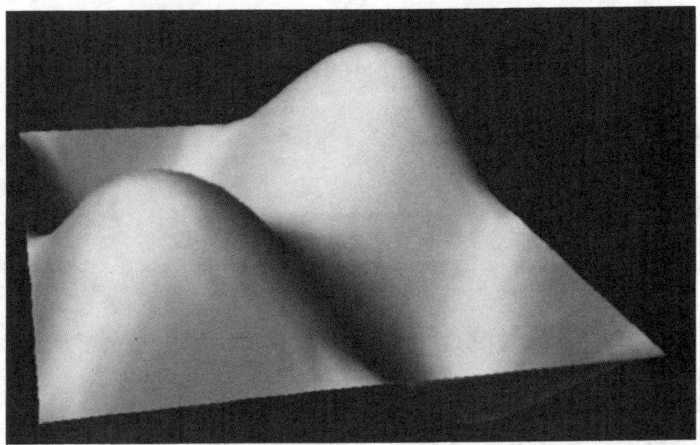

FIG. 9.5.   *The effect of the derivative boundary conditions: Biharmonic PDE surface satisfying boundary conditions (9.6) and (9.10).*

exercise over the surface shape using a combination of function and derivative boundary conditions. We begin with an example of blend generation.

## 9.3.  PDE Surface Design

**9.3.1.  Blend Design.**   Consider Fig. 9.7.  It shows a blend between a cylinder and a flat plane which has been produced by solving (9.4) over the

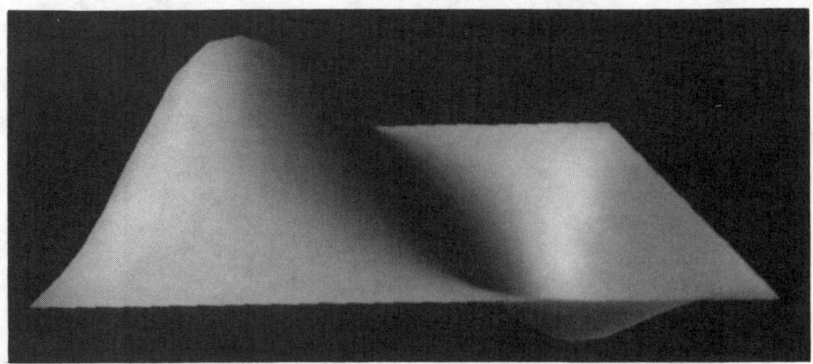

FIG. 9.6.  *The effect of the derivative boundary conditions: Biharmonic PDE surface satisfying boundary conditions (9.6) and (9.11).*

region $[0,1] \times [0,2\pi]$. Although we have considered this blend elsewhere [21], it is useful for the purposes of illustration. Its functional form may be written

$$(9.12) \qquad\qquad x(u,v) = \cosh bu \cos v,$$

$$(9.13) \qquad\qquad y(u,v) = \cosh bu \sin v,$$

$$(9.14) \qquad z(u,v) = (s+3h)(u-1)^2 + (s+2h)(u-1)^3.$$

Note that the blend "trimlines" are the curves

$$(9.15) \qquad\qquad z = h,\ x^2 + y^2 = 1$$

on the cylinder $(u = 0)$, and

$$(9.16) \qquad\qquad z = 0,\ x^2 + y^2 = R^2$$

on the plane $(u = 1)$. The specific boundary conditions on $x, y, z$ and their derivatives are

$$(9.17) \qquad x(0,v) = \cos v, \quad y(0,v) = \sin v, \quad z(0,v) = h,$$

$$(9.18) \qquad x(1,v) = R \cos v, \quad y(1,v) = R \sin v, \quad z(1,v) = 0,$$

where $R = \cosh b$, and

$$(9.19) \qquad x_u(0,v) = 0, \quad y_u(0,v) = 0, \quad z_u(0,v) = s,$$

$$(9.20) \qquad\qquad z_u(1,v) = 0, \quad x_u \text{ and/or } y_u \neq 0.$$

Notice that the blend surface meets the two primary surfaces with continuity of tangent plane. The specific values of the various "shape-parameters" for the surface in Fig. 9.7 are $R = 3$, $h = 2$, $s = -5$, $a = 1/2\pi$. The parameter $s$ controls the magnitude of the vector $\underline{X}_u(0,v)$ around the cylinder trimline; Fig. 9.8, where $s = -12$, shows the effect of increasing its magnitude.

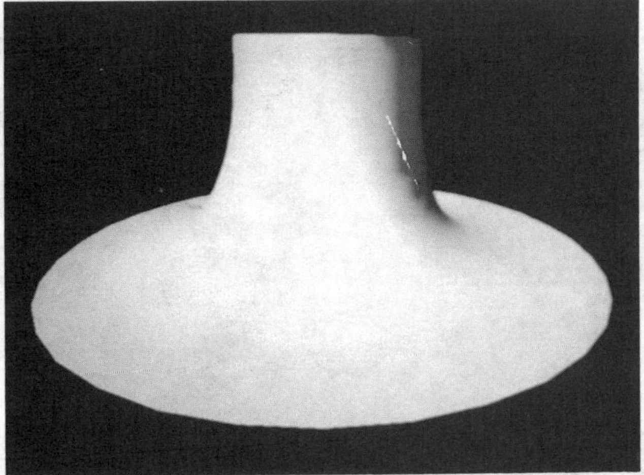

FIG. 9.7.  *Blend surface between cylinder and plane satisfying boundary conditions* (9.17)–(9.20).

FIG. 9.8.  *The effect of increasing the slope parameter s on the surface of Fig. 9.7.*

In these two figures the slope parameter $s$ has been taken to be constant, but this need not be the case. Even within the constraints set by tangent plane continuity, we may allow $s$ to vary and so introduce a degree of design into the creation of a blend. For instance, if we allow $s$ to vary as a function of $v$ thus

$$(9.21) \qquad\qquad s = s_1 + s_2 \cos nv$$

for various choices of $n$, $s_1$, and $s_2$ we can produce a range of interesting shapes, an example of which is shown in Fig. 9.9, which corresponds to the case $s_1 = -5$, $s_2 = 4$, $n = 4$. The application of the method to blend generation is discussed more fully in [1], [21], [22], and [27].

**9.3.2.  Example: Ship Hull.**    In this example we will consider the problem of generating a surface that resembles a ship hull (or rather half a ship hull).

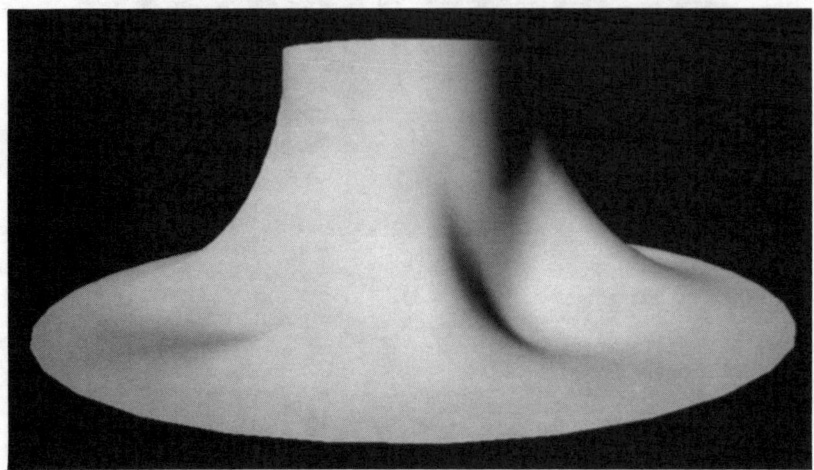

FIG. 9.9.  *The effect of allowing the slope parameters s to vary according to* (9.21).

To do this, (9.4) has been solved over the region $[0,1] \times [0,1]$ using a simple numerical solution technique described in an earlier paper [22]. Figure 9.10 shows the hull designed in this fashion, it consists of two identical half hulls joined together. The boundary conditions on $\underline{X}(u,v)$ were chosen so that the $u = 0$ and 1 isoparametric lines had suitable shapes for the bow and stern, respectively, and so that the $v = 0, 1$ isoparametric lines had suitable shapes for the keel and deck, respectively. The derivative boundary conditions were chosen so that the interior surface resembled a ship hull. Since the solution is numerical, the boundary conditions are imposed discretely at a finite number of points along the appropriate space curves in $E^3$. The mapping between points along the boundary of the $(u, v)$ parameter domain and points in $E^3$ was chosen so that the boundary curves of the hull had arclength parameterization. There was no special reason for this, it was merely a convenient way of ensuring that the numerical resolution was adequate at all points across the surface. As mentioned above, the derivative boundary conditions were chosen so that the interior surface had a geometrically "reasonable" shape. These conditions consisted of specifying $\underline{X}_u$ along the curves $\underline{X}(0,v)$ and $\underline{X}(1,v)$, and specifying $\underline{X}_v$ along the curves $\underline{X}(u,0)$ and $\underline{X}(u,1)$. The magnitudes and directions of the vectors $\underline{X}_u$ and $\underline{X}_v$ were varied until the desired shape was attained.

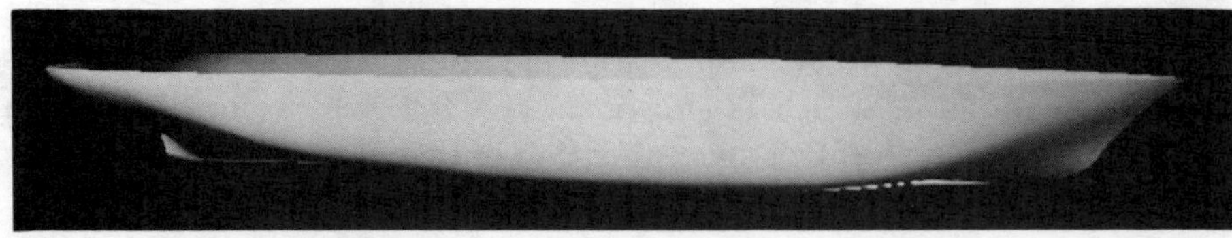

FIG. 9.10.  *PDE ship hull design.*

**9.3.3.  Example: Ship Hull with Chine.**    Figure 9.11 shows a ship hull form in which two chines have been introduced. That is, along each half of the hull there is a curve across which there is an abrupt discontinuity in the direction of surface normal. Such curves are present in some real vessels and is often required for ease of manufacture. The chine has been added to the vessel shown by first designing an initial hull in very much the same fashion as described in the previous example (except that here the hull form is a closed form solution), and then by introducing an internal boundary within the $(u, v)$ parameter space so that $\Omega$ is now divided into two subregions $\Omega_1$ and $\Omega_2$. The solution process is then repeated separately for $\Omega_1$ and $\Omega_2$. The function boundary conditions imposed along the common boundary between $\Omega_1$ and $\Omega_2$ is simply the value of $\underline{X}$ produced by the original solution, while the derivative boundary conditions are chosen so as to produce the appropriate jump in surface normal. For further examples of the application of the PDE method to ship design the reader should consult [23].

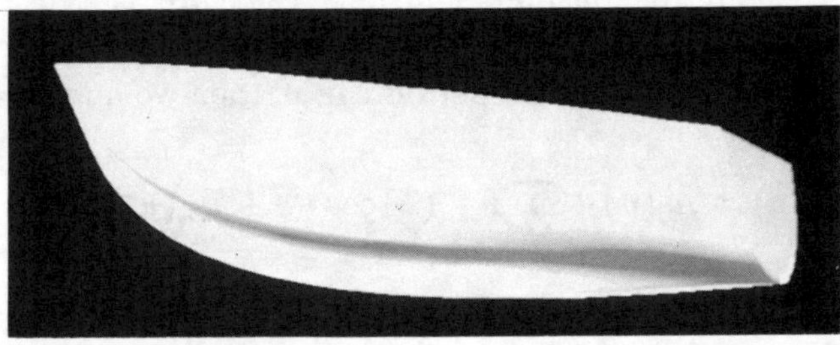

FIG. 9.11.  *PDE ship hull with chines.*

## 9.4.  The Functional Form of a Four-Sided Patch Solution

For the sake of completeness we give in this section a functional form in terms of which any (four-sided) solution to (9.4) may, in principle, be expressed. If we solve equation (9.4) over the domain $[0, \pi] \times [0, \pi]$ in the $(u, v)$ parameter plane (we have chosen this parameter range rather than $[0, 1] \times [0, 1]$ to prevent the proliferation of factors of $\pi$ in the solution), we can write the general form for the solution thus

$$(9.22) \qquad \underline{X}(u, v) = \sum_{n=0}^{\infty} \underline{A}_n(u)\sin(nv) + \underline{B}_n(v)\sin(anu)$$

where the coefficient functions $\underline{A}_n(u)$ and $\underline{B}_n(v)$ are of the form

$$(9.23) \qquad \underline{A}_n(u) = \underline{a}_{n1}e^{anu} + \underline{a}_{n2}ue^{anu} + \underline{a}_{n3}e^{-anu} + \underline{a}_{n4}ue^{-anu},$$

$$(9.24) \qquad \underline{B}_n(v) = \underline{b}_{n1}e^{nv} + \underline{b}_{n2}ve^{nv} + \underline{b}_{n3}e^{-nv} + \underline{b}_{n4}ve^{-nv},$$

and $\underline{a}_{n1}$, $\underline{a}_{n2}$, $\underline{a}_{n3}$, $\underline{a}_{n4}$, and $\underline{b}_{n1}$, $\underline{b}_{n2}$, $\underline{b}_{n3}$, $\underline{b}_{n4}$, are vector-valued constants whose values for a particular solution are determined by the boundary conditions which take the form

$$(9.25) \quad \underline{X}(0,v) = \underline{f}_0(v); \underline{X}(\pi,v) = \underline{f}_1(v); \underline{X}_u(0,v) = \underline{s}_0(v); \underline{X}_u(\pi,v) = \underline{s}_1(v),$$

$$(9.26) \quad \underline{X}(u,0) = \underline{g}_0(u); \underline{X}(u,\pi) = \underline{g}_1(u); \underline{X}_v(u,0) = \underline{t}_0(u); \underline{X}_v(u,\pi) = \underline{t}_1(u).$$

In principle, a Fourier analysis of these boundary conditions may sometimes be used to obtain the coefficient functions $\underline{A}_n(u)$ and $\underline{B}_n(v)$. This has been discussed in more detail elsewhere [16].

## 9.5. Periodic Patches

It is sometimes useful to consider solutions that are closed in one of the parametric directions. For example, if we solve (9.4) over the region $[0,1] \times [0,2\pi]$ and are looking for a solution periodic in $v$, then we may write its general form as

$$(9.27) \qquad \underline{X}(u,v) = \underline{A}_0(u) + \sum_{n=1}^{\infty}(\underline{A}_n(u)\cos nv + \underline{B}_n(u)\sin nv),$$

where

$$(9.28) \qquad \underline{A}_0 = \underline{a}_{00} + \underline{a}_{01}u + \underline{a}_{02}u^2 + \underline{a}_{03}u^3,$$

$$(9.29) \qquad \underline{A}_n = \underline{a}_{n1}e^{anu} + \underline{a}_{n2}ue^{anu} + \underline{a}_{n3}e^{-anu} + \underline{a}_{n4}ue^{-anu},$$

$$(9.30) \qquad \underline{B}_n = \underline{b}_{n1}e^{anu} + \underline{b}_{n2}ue^{anu} + \underline{b}_{n3}e^{-anu} + \underline{b}_{n4}ue^{-anu},$$

where $\underline{a}_{n1}$, $\underline{a}_{n2}$, $\underline{a}_{n3}$, $\underline{a}_{n4}$, and $\underline{b}_{n1}$, $\underline{b}_{n2}$, $\underline{b}_{n3}$, $\underline{b}_{n4}$, are vector-valued constants, determined by the boundary conditions imposed on the isoparametric lines $u = 0$ and $u = 1$. As above, these conditions are of the form

$$(9.31) \qquad \begin{aligned} \underline{X}(0,v) &= \underline{f}_0(v); & \underline{X}(1,v) &= \underline{f}_1(v); \\ \underline{X}_u(0,v) &= \underline{s}_0(v); & \underline{X}_u(1,v) &= \underline{s}_1(v). \end{aligned}$$

The functions $\underline{f}_0(v)$ and $\underline{f}_1(v)$ give the shapes of the curves ($u = 0, 1$) bounding the surface, and, if the surface is closed in the $v$ direction, they must satisfy the condition

$$(9.32) \qquad \underline{f}_0(0) = \underline{f}_0(2\pi), \quad \underline{f}_1(0) = \underline{f}_1(2\pi).$$

As in the case of the four-sided patch, the coefficient functions $\underline{A}_n(u)$ and $\underline{B}_n(v)$ can, in principle, be obtained from the boundary conditions by Fourier analysis, but such an approach is of limited use in the general case, where a numerical solution would be more appropriate. However, as has been demonstrated elsewhere [16], closed-form solutions can be given for a number of surfaces of practical significance, the blend of 9.3.1 being a surface of this type. In the sections that follow, further examples of periodic solutions are given.

**9.5.1. Example: Yacht Hull Design.**   Figure 9.12 shows a yacht hull designed from a periodic PDE patch. The boundary conditions have been chosen so that the curve $\underline{X}(0,v)$ gives the outline of the deck planform; in this case it is the curve defined by

$$x = b_1 \cos v, \quad y = \sin 2v + b_4 \sin 4v + b_6 \sin 6v, \quad z = 0, \quad -\pi/2 \le v \le \pi/2$$
(9.33)

and this is the boundary condition on $\underline{X}(0,v)$. Similarly, the curve $\underline{X}(1,v)$ gives the outline of the bottom of the keel and this is taken to be

$$(9.34) \quad x = x_f + a_f \cos 2v, \quad y = b_f \sin 2v, \quad z = -d, \quad -\pi/2 \le v < \pi/2.$$

For the surface described in this section the various constants in (9.33) and (9.34) are as follows: $b_1 = 6$, $b_4 = -0.1$, $b_6 = -0.06$, $d = 1.5$, $x_f = 3.7$, $a_f = 0.5$, $b_f = 0.05$. Note that the range in $v$ has been reduced, so that the singular nature of the surface at the tip of the sharp bow can be accounted for in the closed-form solution without the introduction of Fourier modes with noninteger coefficients. These boundary conditions suggest that the most simple solution we can look for is one of the form

$$(9.35) \qquad \underline{X}(u,v) = \underline{A}_0(u) + \sum_{n=1}^{6}(\underline{A}_n(u)\cos nv + \underline{B}_n(u)\sin nv),$$

where the only nonzero vector coefficients are

$$
(9.36) \quad
\underline{A}_0 = \begin{Bmatrix} X_0(u) \\ 0 \\ Z_0(u) \end{Bmatrix}, \quad
\underline{A}_1 = \begin{Bmatrix} X_1(u) \\ 0 \\ 0 \end{Bmatrix}, \quad
\underline{A}_2 = \begin{Bmatrix} X_2(u) \\ 0 \\ 0 \end{Bmatrix},
$$

$$
\underline{B}_2 = \begin{Bmatrix} 0 \\ Y_2(u) \\ 0 \end{Bmatrix}, \quad
\underline{B}_4 = \begin{Bmatrix} 0 \\ Y_4(u) \\ 0 \end{Bmatrix}, \quad
\underline{B}_6 = \begin{Bmatrix} 0 \\ Y_6(u) \\ 0 \end{Bmatrix},
$$

The derivative boundary conditions $\underline{X}_u(0,v)$ and $\underline{X}_u(1,v)$ are restricted by this choice of solution, but realistic derivative conditions can be applied on the bounding curves $u = 0$ and $u = 1$, so that an appropriate surface shape for a yacht is obtained. At the level of the deck $(u = 0)$, the following boundary conditions have been applied

$$(9.37) \qquad \underline{X}_u(0,v) = \begin{pmatrix} -sX_0 \\ 0 \\ -sZ_0 \end{pmatrix},$$

where $sX_0$ and $sZ_0$ are positive constants. These conditions imply that the isoparametric $v$ lines leave the boundary curve $u = 0$ at an angle of $\tan^{-1}(sX_0/sZ_0)$. On $u = 1$, the boundary curve at the bottom of the boat, to obtain a yacht-like hull, we have taken

$$(9.38) \qquad \underline{X}_u(1,v) = \begin{pmatrix} -sX_f \\ 0 \\ -sZ_f \end{pmatrix}.$$

This choice of derivative conditions ensures that we get a fin-like shape for the keel, providing $sZ_f$ is sufficiently large and positive and further the keel is raked toward the stern at an angle of $\tan^{-1}(sX_f/sZ_f)$ to the vertical. Note that if further design is required, the means by which this can be achieved, in particular by a modification of the derivative boundary conditions, has been discussed in detail elsewhere [16], [23], [26].

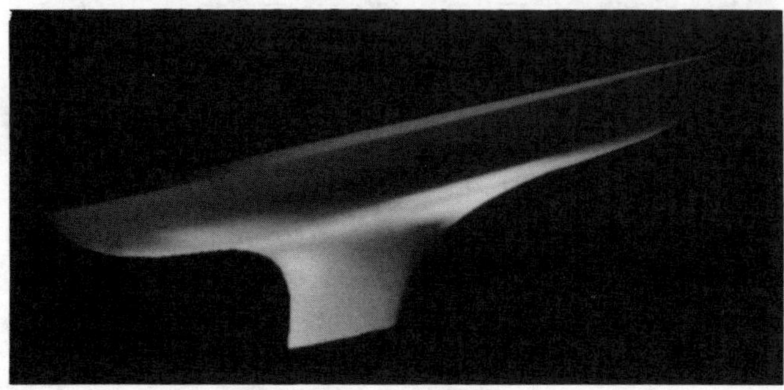

Fig. 9.12.  *Yacht hull designed from a closed-form periodic solution.*

**9.5.2.  Example:  SWATH Hull Design.**  With a slight modification to the boundary conditions we can produce the object in Fig. 9.13, which represents the submerged hull of a SWATH (Small Waterplane Area Twin Hull) vessel. SWATH [24] is a ship design that basically resembles a platform raised clear of the sea surface by twin slender struts. At the end of these struts, submerged below the waterline, is an object resembling that shown in Fig. 9.13. Note that to give a better indication of the relationship of the hull to the rest of the vessel, part of the strut that connects to the working platform of the vessel is also shown. Although the functional form of the SWATH hull is basically that of the yacht hull, to control the curvature at the keel a sixth-order PDE solution for $z$ was required, i.e., $\Delta^2 x, y = 0 \; \Delta^3 z = 0$. Now, the function boundary conditions are such that the curve $\underline{X}(0, v)$ gives the curve along which the hull meets the supporting strut, while the curve $\underline{X}(1, v)$ has been shrunk to a point. The derivative boundary conditions $\underline{X}_u(0, v)$ and $\underline{X}_u(1, v)$ have also been modified so that an appropriate shape for the SWATH hull is obtained; for instance, around the curve $\underline{X}(1, v)$, their effect is to "inflate" the surface.

**9.5.3.  Example: Propeller Design.**  A propeller blade is another object that can be derived from the basic yacht solution, this time by twisting and offsetting the deck relative to the keel, and then by shrinking the yacht's keel to a point so that there is no longer a "hole" in the surface at this point. Figure 9.14 shows an assemblage of such blades in a marine propeller.

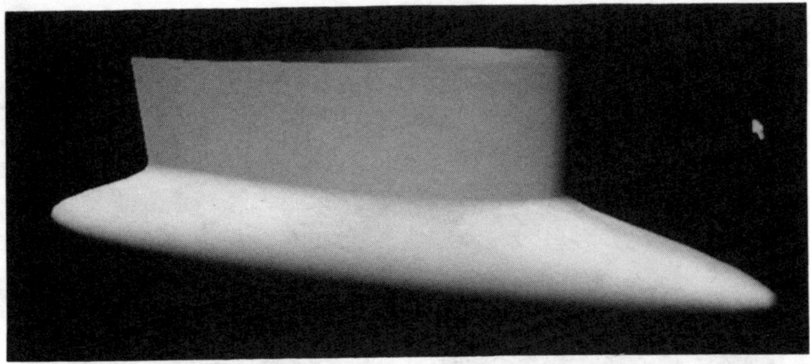

FIG. 9.13. *SWATH hull design.*

FIG. 9.14. *Propeller design: each blade is fashioned from a PDE surface.*

**9.5.4. Phone Handset Design.** As yet another example of how a fourth-order blend solution can be used to produce a free-form surface, consider Fig. 9.15. The handset is constructed from four closed-form surface patches each of which is a modification of the basic blend shape: two patches for the main "body" of the handset, a patch for the mouthpiece, and a patch for the earpiece.

## 9.6. Conclusions

We have tried in this paper to indicate the range of surface shapes that are accessible to this boundary-value approach. In many cases it is possible to

                                                                    

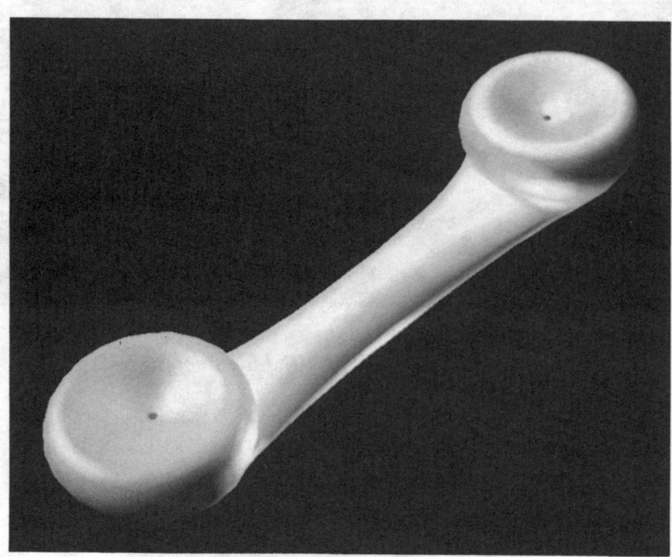

FIG. 9.15.  *PDE phone handset.*

create much of an object's surface from a single PDE patch, and even in those
cases where the object is so complicated that this is not possible, one can
divide the surface into a "quilt" of such patches as necessary.  Although we
have tended to work with a mathematical formulation of the problem, it is
our belief that this should not be necessary for a designer.  Our experience of
designing objects with the method indicates that PDE surfaces behave in a
geometrically intuitive fashion under the influence of the boundary conditions.
Thus one could imagine design taking place interactively at the screen of a
suitable workstation.  To create the desired shape the designer would first
specify the shape of the curves bounding the surface in space, then "flesh-
out" the surface by giving the derivative conditions, perhaps by specifying the
appropriate "vector-field" around the bounding curves.

Where this method fundamentally differs from most conventional approaches to surface design is in its boundary value approach, which implies
that surfaces are specified by data distributed along curves, rather than across
surfaces in the form of a "control-mesh."  This means that, in terms of the
number of parameters required to specify a surface, the "dimensionality" of
the problem has been reduced from two to one.  This may not be particularly
important if one is solely interested in an object as a geometrical entity, but
for any object whose physical interaction with its environment is important
for its intended use, the number of "shape-parameters" required to describe it
may be crucial.  This is because it is desirable that its shape be in some sense
optimized with respect to its functionality, and to achieve this it is necessary
to calculate the relevant physical properties of a given shape and how these

properties change with alterations in the shape parameters. Now, since the calculation of the object's physical characteristics is often a numerically intensive operation, the more shape parameters there are the greater the costs of any shape optimization, hence the value of any technique that tends to reduce the number of shape parameters. For the application of the present method to functional shape design the reader should consult [25] and [26].

## Acknowledgments

The authors wish to thank the SERC for financial support.

## References

[1] M. I. G Bloor and M. J. Wilson, *Generating blend surfaces using partial differential equations*, Comput. Aided Des., 21 (1989), pp. 165–171.

[2] I. D. Faux and M. J. Pratt, *Computational Geometry for Design and Manufacture*, Ellis Horwood, Chichester, 1979.

[3] M. E. Mortenson, *Geometric Modeling*, Wiley-Interscience, New York, 1985.

[4] C. D. Woodward, *Cross-sectional design of B-spline surfaces*, Comput. Graphics, 11 (1987), pp. 193–201.

[5] W. Tiller, *Rational B-splines for curve and surface representation*, IEEE Comput. Graph. Appl., 3 (1983), pp. 61–69.

[6] L. Piegl and W. Tiller, *Curve and surface constructions using rational B-splines*, Comput. Aided Des., 19 (1987), pp. 485–498.

[7] T. Várady, *An experimental system for interactive design and manufacture of sculptured surfaces*, Computers in Industry, 3 (1982), pp. 125–135.

[8] C. D. Woodward, *Skinning techniques for interactive B-spline surface interpolation*, Comput. Aided Des., 20 (1988), pp. 441–451.

[9] M. I. G. Bloor and M. J. Wilson, *Representing PDE surfaces in terms of B-splines*, Comput. Aided Des., 22 (1990), pp. 324–331.

[10] J. M. Brown, M. I. G. Bloor, M. S. Bloor, and M. J. Wilson, *Generating B-spline representations of PDE surfaces using the finite-element method*, Comput. Aided Geom. Des., 1990, submitted.

[11] ______, *Generation and modification of non-uniform B-spline surface approximations to PDE surfaces using the finite-element method*, in Advances in Design Automation, Vol. 1: Computer Aided and Computational Design, B. Ravani, ed., ASME, 1990, pp. 265–272.

[12] G. Celniker and D. Gossard, *Energy-based models for free-form surface shape design*, ASME Design Automation Conference, 1988, pp. 107–112.

[13] H. Hagen and G. Schulze, *Automatic smoothing with geometric surface patches*, Comput. Aided Geom. Des., 4 (1987), pp. 231–235.

[14] H. Nowacki, L. Dingyuan, and l. Xinmin, *Mesh fairing $GC^1$ surface generation method*, in Theory and Practise of Geometric Modeling, W. Straßer and H.-P. Seidel, eds., Springer-Verlag, Berlin, 1989, pp. 93–103.

[15] C. J. K. Williams, *Use of structural analogy in generation of smooth surfaces for engineering purposes*, Comput. Aided Des., 19 (1987), pp. 310–322.

[16] M. I. G. Bloor and M. J. Wilson, *Using partial differential equations of generate free-form surfaces*, Comput. Aided Des., 22 (1990), pp. 202–212.

[17] ______, *Local Control of Surfaces Generated using PDEs*, in preparation.

[18] J. J. Stoker, *Differential Geometry*, Wiley-Interscience, New York, 1969.

[19] A. P. Rockwood, *Introducing Sculptured Surfaces into a Geometric Modeler*, in Solid Modeling by Computer: From Theory to Applications, M. S. Pickett and J. W. Boyse, eds., Plenum Press, London, 1984, pp. 237–258.

[20] E. Zauderer, *Partial Differential Equations of Applied Mathematics*, Wiley-Interscience, New York, 1983.

[21] M. I. G. Bloor and M. J. Wilson, *Blend design as a boundary-value problem*, W. Straßer and H.-P. Seidel, eds., in Theory and Practise of Geometric Modeling, Springer-Verlag, Berlin, 1989, pp. 221–234.

[22] ______, *Generating N-sided patches with partial differential equations*, in Recent Advances in Computer Graphics, R. A. Earnshaw and B. Wyvill, eds., Springer-Verlag, Berlin, 1989, pp. 129–145.

[23] ______, *Geometric design of hull forms using partial differential equations*, in CFD and CAD in Ship Design, G. van Oortmeressen, ed., Elsevier, Amsterdam, 1990.

[24] J. R. MacGregor, R. R. Simpson, and P. Norton, *Parametric Studies in the Design of a Series of Swath Ships*, Intersociety Advanced Marine Vehicles Conference, AIAA-89-1476, 1989.

[25] T. W. Lowe, M. I. G. Bloor, and M. J. Wilson, *Functionality in blend design*, Comput. Aided Des., November 1990.

[26] ______, *Functionality in surface design*, in Advances in Design Automation, Vol. 1: Computer Aided and Computational Design, B. Ravani, ed., ASME, 1990, pp. 43–50.

[27] S. Y. Cheng, M. I. G. Bloor, A. Saia, and M. J. Wilson, *Blending between quadric solids using partial differential equations*, in Advances in Design Automation, Vol. 1: Computer Aided and Computational Design, B. Ravani, ed., ASME, 1990, pp. 257–263.

# Modeling with Box Spline Surfaces

Morten Dæhlen

## 10.1. Introduction

On a three-direction grid in $\mathrm{I\!R}^2$ we will construct surfaces that are linear combinations of translates of box splines. Box splines on a three-direction grid are piecewise polynomials defined over triangles. This gives us the opportunity to define nonrectangular surfaces, which often arise in computer aided geometric design (CAGD). With special emphasis on three-, five-, and six-sided surfaces we will consider some important aspects related to an embedding of box spline surfaces within a tensor product framework.

Box splines are generalizations of tensor products to nonrectangular meshes, and much of the univariate machinery for manipulation of B-spline series can be generalized. They are piecewise polynomials; translates of box splines on a three-direction grid are linearly independent; they form a partition of unity, they are locally supported, etc. Moreover, knotline insertion or subdivision algorithms for box splines are well known. Such algorithms play a key role in CAGD and are the basis for efficient rendering of spline surfaces, computation of intersections, etc.

We are going to study box spline surfaces defined on finite regions in $\mathrm{I\!R}^2$. The three-direction grid and the regions of interest are defined in §10.2. To imbed box spline surfaces within a tensor product framework, we need to match the two surface types along common boundaries. This is the topic of §10.3. In §10.4 we propose a method that uses box splines for the construction of nonrectangular box spline surfaces in regions where nonrectangular surfaces naturally arise when fillets and blends are constructed between tensor product surfaces.

Although it is not crucial, we assume that the reader is familiar with univariate B-splines and tensor product B-spline surfaces ([2],[24]).

## 10.2. Box Splines

A systematic treatment of bivariate box splines on a three-direction mesh was first given in [23]. The general multivariate definition was introduced in [3] and further studied in [4]. A study of box splines defined on a three-direction grid in $\mathbb{R}^2$ was originally given in [5]. For surveys of these and other results, see [14],[15]. Since translates of box splines on a three-direction grid are linearly independent ([17],[18],[20]), it is natural to use these functions as a basis for the construction of surfaces. Results on box splines defined on other meshes can be found in [7],[14], and references therein.

Given two linearly independent vectors $\mathbf{c}^1$ and $\mathbf{c}^2$ in $\mathbb{R}^2$, the three-direction grid is constructed by drawing straight lines in the three directions $\mathbf{c}^1$, $\mathbf{c}^2$, and $\mathbf{c}^1 + \mathbf{c}^2$ through all points in $\mathbb{R}^2$ of the form $j\mathbf{c}^1 + k\mathbf{c}^2$, $j, k \in \mathbb{Z}$. We will consider the standard mesh $G$ obtained by choosing $\mathbf{c}^1 = \mathbf{d}^1 = (1,0)^T$ and $\mathbf{c}^2 = \mathbf{d}^2 = (0,1)^T$. Hence, $\mathbf{d}^3 = (1,1)^T$ is the third direction.

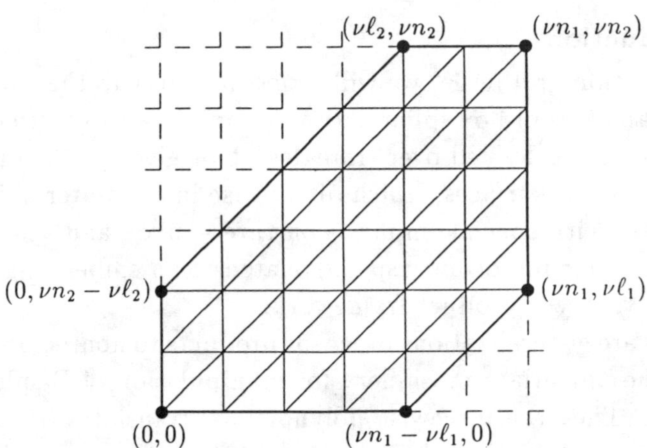


FIG. 10.1. *A bounded, polygonal convex set $\Omega$ with boundaries along the three directions in the grid.*

Let $\Omega$ be a convex subset of $\mathbb{R}^2$ with an interior $\Omega^0$. We assume that the boundary of $\Omega$ consists of mesh lines, and $\Omega$ is obtained by removing two triangles of size $\nu\ell_1$ and $\nu\ell_2$ from the lower right and upper left corner of a rectangle of size $\nu n_1 \times \nu n_2$. We require $\nu n_1 > 0$, $\nu n_2 > 0$, and $0 \le \nu\ell_1, \nu\ell_2 \le \min(\nu n_1, \nu n_2)$. Depending on the values of $n_1$, $n_2$, $\ell_1$, and $\ell_2$ we obtain a triangle, a trapezoid, a pentagon and a hexagon. Figure 10.1 shows a convex set $\Omega$ given by the integers $n_1$, $n_2$, $\ell_1$, $\ell_2$, and the scaling factor $\nu$.

Let $\mathbf{x} = (x,y) \in \mathbb{R}^2$. Given three integers $m_1$, $m_2$ and $m_3$ with $m_1 > 0$, $m_2 > 0$, and $m_3 \ge 0$, we define box splines on the grid $G$ as follows:

$$B^{\mathbf{m}}(\mathbf{x}) = B^{(m_1,m_2,m_3)}(x,y)$$

$$(10.1) \qquad = \begin{cases} M_{m_1}(x)M_{m_2}(y) & \text{if } m_3 = 0, \\ \int_0^1 B^{(m_1,m_2,m_3-1)}(x-t, y-t)\,dt, & \text{otherwise.} \end{cases}$$

Here, $M_k(t) = M(t|0, 1, \cdots, k)$ is the univariate B-spline of order $k$ with knots $0, 1, \cdots, k$ normalized to have unit integral. Thus, $B^{\mathbf{m}}$ is a tensor product B-spline if $m_3 = 0$.

Equation (10.1) defines $B^{\mathbf{m}}$ for every point in $\mathbb{R}^2$. If we define the univariate B-spline to be right continuous, i.e., we set

$$M_1(t) = \chi_{[0,1)}(t) = \begin{cases} 1, & \text{if } 0 \le t < 1, \\ 0, & \text{otherwise,} \end{cases}$$

then the value of $B^{\mathbf{m}}$ or one of its derivatives on a grid line is obtained as the limit from the right in the $x$-direction and from above in the $y$-direction.

We will recall some properties of box splines. Set $| \mathbf{m} |= m_1 + m_2 + m_3$. The box spline $B^{\mathbf{m}}$ is a piecewise polynomial of degree $| \mathbf{m} | - 2$ on each triangle of $G$. The support of $B^{\mathbf{m}}$ is a six-sided polygonal region given by

$$\operatorname{supp} B^{\mathbf{m}} = \sum_{i=1}^{3} t_i \mathbf{d}^i, \quad 0 \le t_i \le m_i, \ i = 1, 2, 3.$$

Moreover, partial derivatives of order $\le | \mathbf{m} | - 2 - \max m_j$, $j = 1, 2, 3$ are continuous everywhere.

Let $D_{\mathbf{y}}$ denote the directional derivative in the direction $\mathbf{y}$ in $\mathbb{R}^2$. The directional derivative of a box spline is given by the following recurrence relation:

$$D_{\mathbf{y}} B^{\mathbf{m}}(\mathbf{x}) = \sum_{q=1}^{3} \lambda_q \{ B^{\mathbf{m}^q}(\mathbf{x}) - B^{\mathbf{m}^q}(\mathbf{x} - \mathbf{d}^q) \},$$

where $\mathbf{y} = \sum_{q=1}^{3} \lambda_q \mathbf{d}^q$ and $\mathbf{m}^1 = (m_1 - 1, m_2, m_3)$, $\mathbf{m}^2 = (m_1, m_2 - 1, m_3)$, and $\mathbf{m}^3 = (m_1, m_2, m_3 - 1)$. If $\mathbf{y} = \mathbf{d}^j$, $j \in \{1, 2, 3\}$ then

$$(10.2) \qquad D_{\mathbf{d}_j} B^{\mathbf{m}}(\mathbf{x}) = B^{\mathbf{m}^j}(\mathbf{x}) - B^{\mathbf{m}^j}(\mathbf{x} - \mathbf{d}^j).$$

In order to define box spline surfaces as linear combinations of translates of box splines, let $\Omega$ be a region in $\mathbb{R}^2$ bounded by straight lines in the three-direction grid $G/\nu$. Given $\mathbf{m} = (m_1, m_2, m_3)$, the family of box splines surfaces on $\Omega$ we denote by

$$S_{\mathbf{m}, \nu}(\Omega) = \left\{ \sum_{\mathbf{i} \in I_{\mathbf{m}, \nu}(\Omega)} c_{\mathbf{i}} B^{\mathbf{m}}(\nu \mathbf{x} - \mathbf{i}) : c_{\mathbf{i}} \in \mathbf{R} \right\}$$

where

$$I_{\mathbf{m}, \nu}(\Omega) = \{ \mathbf{i} \in \mathbb{Z}^2 : B^{\mathbf{m}}(\nu \mathbf{x} - \mathbf{i}) \ne 0 \quad \text{for some} \quad \mathbf{x} \in \Omega^0 \}.$$

Here the box splines $B^{\mathbf{m}}(\nu \mathbf{x} - \mathbf{i})$ are translates on $\mathbb{Z}^2/\nu$. The index set $I_{\mathbf{m}, \nu}(\Omega)$ indicates the lower left corner of the support of the translates of the box splines $B^{\mathbf{m}}(\nu \mathbf{x} - \mathbf{i})$ whose support overlaps $\Omega$.

Since translates of box splines on a three-direction mesh are linearly independent ([17],[18],[20]), the dimension of the spline space $S_{\mathbf{m},\nu}(\Omega)$ is equal to the number of elements in $I_{\mathbf{m},\nu}(\Omega)$. From [11] we have

$$
\begin{aligned}
N =\mid I_{\mathbf{m},\nu}(\Omega) \mid &= (\nu n_1 + m_1 + m_3 - 1)(\nu n_2 + m_2 + m_3 - 1) \\
&\quad - \frac{(\nu \ell_1 + m_3 - 1)(\nu \ell_1 + m_3)}{2} \\
&\quad - \frac{(\nu \ell_2 + m_3 - 1)(\nu \ell_2 + m_3)}{2}.
\end{aligned}
$$
(10.3)

By allowing the coefficients $c_{\mathbf{i}}$, $\mathbf{i} \in I_{\mathbf{m},\nu}(\Omega)$ to be vectors in $\mathbb{R}^d, d \geq 1$, we have a parametric box spline surface in the Eucledian space $\mathbb{R}^d$ given by

$$
\mathbf{f}(\mathbf{x}) = \begin{pmatrix} f_1(\mathbf{x}) \\ \vdots \\ f_d(\mathbf{x}) \end{pmatrix} = \sum_{\mathbf{i} \in I_{\mathbf{m},\nu}(\Omega)} c_{\mathbf{i}} B^{\mathbf{m}}(\nu \mathbf{x} - \mathbf{i}).
$$

In the following we assume that $d = 3$, that is, we are going to study parametric box spline surfaces in $\mathbb{R}^3$. To do this, we need methods for the construction of such surfaces. A well-known method in geometric modeling is to manipulate the coefficients (control points) $c_{\mathbf{i}}$ until you are satisfied with the shape of the surface (see [22]). A method for grid point interpolation with smoothing and a method based on least squares are described in [1] and [13], respectively. A survey on these and other results can be found in [14].

Since knotline insertion and line averaging algorithms are important in CAGD, we conclude this section by recalling some details. Line averaging, subdivision, and knotline insertion for box splines are covered in detail in [6],[8],[14],[16].

Consider the spaces $S_{\mathbf{m}}(\Omega) = S_{\mathbf{m},1}(\Omega)$ and $S_{\mathbf{m},\nu}(\Omega)$. For $\nu > 1$ the grid $G/\nu$ corresponding to $S_{\mathbf{m},\nu}(\Omega)$ will be a refinement of the grid $G$ corresponding to $S_{\mathbf{m}}(\Omega)$.

From [8] we recall that

$$
(10.4) \qquad B^{\mathbf{m}}(\mathbf{x}) = \sum_{\mathbf{i} \in \mathbb{Z}^2} \beta_{\mathbf{m}}^{\nu}(\mathbf{i}) B^{\mathbf{m}}(\nu \mathbf{x} - \mathbf{i}), \quad \mathbf{x} \in \mathbb{R}^2,
$$

where the $\beta$s are given as the coefficients to a function, which takes the following simple form

$$
\begin{aligned}
q_{\mathbf{m}}^{\nu}(\mathbf{u}) = q_{\mathbf{m}}^{\nu}(u,v) &= \sum_{\mathbf{i}=(i,j)} \beta_{\mathbf{m}}^{\nu}(\mathbf{i}) u^i v^j \\
&= \nu^{2-|\mathbf{m}|}(1 + u)^{m_1}(1 + v)^{m_2}(1 + uv)^{m_3}.
\end{aligned}
$$

We call $\{\mathbf{i}, \beta_{\mathbf{m}}^{\nu}(\mathbf{i})\}$ a mask for $B^{\mathbf{m}}$.

Suppose $\mathbf{f} \in S_{\mathbf{m}}$. It then follows that $\mathbf{f} \in S_{\mathbf{m},\nu}$ for all $\nu \in \mathbb{N}$. Therefore,

$$
\mathbf{f}(\mathbf{x}) = \sum_{\mathbf{j}} c_{\mathbf{j}} B^{\mathbf{m}}(\mathbf{x} - \mathbf{j}) = \sum_{\mathbf{i}} b_{\mathbf{i}} B^{\mathbf{m}}(\nu \mathbf{x} - \mathbf{i}),
$$

for certain coefficients $c_j$ and $b_i$. Using (10.4) we obtain a relation between the two sets of coefficients. Indeed, since

$$\mathbf{f}(\mathbf{x}) = \sum_j c_j \sum_i \beta_{\mathbf{m}}^{\nu}(\mathbf{i}) B^{\mathbf{m}}(\nu(\mathbf{x} - \mathbf{j}) - \mathbf{i})$$

we find after rearranging sums that

$$b_{\mathbf{i}} = \sum_j \beta_{\mathbf{m}}^{\nu}(\mathbf{i} - \nu\mathbf{j})c_{\mathbf{j}}.$$

A stable line averaging algorithm for computing the $b_i$s from the $c_j$s can be found in [6],[14]. All surface examples in this text are accomplished by using the line averaging algorithm, together with the box spline evaluation method introduced in [10]. For more details on box spline basics and multivariate splines, see [7],[14],[15],[19].

## 10.3.  Matching Edges

The use of box spline surfaces together with tensor product B-spline surfaces requires a discussion on problems related to the construction of joints between the two surface types. We will consider the problem of extending a given box spline surface with tensor product B-spline surfaces. The converse is in general not possible since box splines are defined on uniform partitions and tensor products in general are defined on nonuniform partitions. We will therefore consider the following problems:

- The construction of a tensor product surface that exactly matches a partial derivative along an edge of a box spline surface.

- The construction of a tensor product surface that matches a linear combination of partial derivatives along an edge of a box spline surface. In the literature this is referred to as tangent plane continuity.

- The construction of an overall tangent plane continuous surface consisting of tensor product surfaces, which completely surround a given box spline surface.

Our discussion here is concerned with the construction of curves and surfaces along the edges of a given box spline surface. We will not cover problems, where additional assumptions are given in the construction of the surrounding tensor product surfaces.

The construction of a tensor product surface which is $C^0$ with a given box spline surface is trivial, since edge curves of box spline surfaces are piecewise polynomials and therefore B-spline curves.

The derivatives of a box spline basis function are given in (10.2). Given a parametric box spline surface

$$\mathbf{f}(x) = \sum_{\mathbf{p} \in I_{\mathbf{m}}(\Omega)} c_{\mathbf{p}} B_{\mathbf{p}}^{\mathbf{m}}(\mathbf{x}), \quad \mathbf{x} \in \mathbb{R}^2,$$

we have the directional derivatives

$$D_{\mathbf{d}^i}\mathbf{f}(\mathbf{x}) = \sum_{\mathbf{p}\in I_{\mathbf{m}^i}(\Omega)} (c_{\mathbf{p}} - c_{\mathbf{p}-\mathbf{d}^i})B_{\mathbf{p}}^{\mathbf{m}^i}(\mathbf{x}), \quad \mathbf{x}\in \mathbb{R}^2,\ i = 1, 2, 3\ ,$$

where $\mathbf{m}^1 = (m_1-1, m_2, m_3)$, $\mathbf{m}^2 = (m_1, m_2-1, m_3)$ or $\mathbf{m}^3 = (m_1, m_2, m_3-1)$. The derivative of a box spline surface of degree $|\mathbf{m}| - 2$ is another box spline surface of degree $|\mathbf{m}| - 3$. The directional derivatives of the $C^1$-cubic box spline surface based on $B^{(2,2,1)}$ in the directions $\mathbf{d}^1$ and $\mathbf{d}^2$ are $C^0$-quadratic box spline surfaces based on $B^{(1,2,1)}$ and $B^{(2,1,1)}$, respectively. In the direction $\mathbf{d}^3 = (1,1)^T$ the derivative surface is a bilinear tensor product surface based on $B^{(2,2,0)}$. This implies that cross-boundary partial derivatives of a $C^1$-cubic box spline surface are the $C^0$ piecewise quadratic polynomials.

We recall that cross-boundary partial derivatives of a tensor product surface are of the same degree and continuity as the boundary curve itself. Thus, to produce a tangent continuous match between a box spline surface and a tensor product B-spline surface we have to represent the boundary curve and the cross-boundary derivatives from the box spline surface on the same knot vector.

To explain this, let a $C^2$-quartic box spline surface be given. In this case the boundary curve is $C^2$-quartic and the cross-boundary partial derivatives are $C^1$-cubic. The $C^2$-quartic boundary curve can be written as a degree four B-spline curve with double knots. By inserting one knot at each double knot we obtain a degree four B-spline curve with triple knots. The $C^1$-cubic boundary curves of the cross-boundary derivative box spline surfaces can be written as degree three B-spline curves with double knots. By raising the degree ([9]) of the derivative curves we obtain degree four B-spline curves with triple knots. The boundary curve and the cross-boundary curves now have the same degree and are represented on the same knot vector. Thus, we can construct a continuous match between the box spline surface and the tensor product B-spline surface.

The construction of tensor product B-spline surfaces that exactly match a partial derivative of box spline surfaces is in many cases not very useful. This can be observed, for example, in Fig. 10.8. The tensor product surfaces in Fig. 10.8 have to match two different partial derivatives along the edges of the triangular box spline surface. In particular, the tensor product B-spline surfaces extending the triangular box spline surface along one edge have to match one partial derivative at one end of the edge and another partial derivative at the other end of the edge. We will therefore discuss the construction of composite cross-boundary derivatives.

Let $\Omega$ be a domain in $\mathbb{R}^2$ such that the boundary of $\Omega$ consists of lines in the three-direction grid. It follows that cross-boundary derivatives at a particular edge of the surface can be constructed as a linear combination of the partial derivatives in the two directions not parallel to the gridline along the edge. The edge of interest are parallel to one of the three vectors $\mathbf{d}^1 = (1,0)^T$, $\mathbf{d}^2 =$

$(0,1)^T$, or $\mathbf{d}^3 = (1,1)^T$. We want to construct a composite cross-boundary derivative given by the two directions not parallel to the boundary. We denote these two directions by $\mathbf{u}$ and $\mathbf{v}$, where $\mathbf{u}, \mathbf{v} \in \{\mathbf{d}^1, \mathbf{d}^2, \mathbf{d}^3\}$, $\mathbf{u} \neq \mathbf{v}$. Let $\lambda_{\mathbf{u}}(t)$, $t \in [0,1]$ be the B-spline boundary curve of the derivative box spline surface $D_{\mathbf{u}}\mathbf{f}$, and $\lambda_{\mathbf{v}}(t)$, $t \in [0,1]$ the B-spline boundary curve of the derivative box spline surface $D_{\mathbf{v}}\mathbf{f}$. Without loss of generality we let $\lambda_{\mathbf{u}}$ and $\lambda_{\mathbf{v}}$ be B-spline curves defined on the same partition of the interval $[0,1]$. Now, let

$$h(t) = w_{\mathbf{u}}(t)\lambda_{\mathbf{u}}(t) + w_{\mathbf{v}}(t)\lambda_{\mathbf{v}}(t),$$

where $w_{\mathbf{u}}(t)$ and $w_{\mathbf{v}}(t)$ are suitable weight functions on the interval $[0,1]$. If $w_{\mathbf{u}}(t)$ and $w_{\mathbf{v}}(t)$ are polynomials of degree $\leq n$, it follows that $h(t)$ is a B-spline curve of degree $|\mathbf{m}| - 3 + n$, since $\lambda_{\mathbf{u}}$ and $\lambda_{\mathbf{v}}$ are degree $|\mathbf{m}| - 3$ B-spline curves. With $w_{\mathbf{u}}(t) = (1 - t)$ and $w_{\mathbf{v}}(t) = 1 - w_{\mathbf{u}}(t)$ we have that $h(t)$ is a degree $|\mathbf{m}| - 2$ B-spline curve, where $h(0) = \lambda_{\mathbf{u}}$ and $h(1) = \lambda_{\mathbf{v}}$.

Since $h(t)$ is no longer a strictly partial derivative, but lies in the tangent plane of the surface $\mathbf{f}$ along the actual edge, we obtain a tangent plane continuous match between the box spline surface and the tensor product B-spline surface.

To completely surround a box spline surface with tensor product B-spline surfaces, such that the overall surface becomes tangent plane continuous, we have to take care of the mixed partial derivatives, known as the twist vectors. At one corner of a box spline surface we have to attach three tensor product B-spline surfaces, two surfaces matching along each edge attached to the corner, and one surface that has the corner point in common with the box spline surface (see Figs. 10.6–10.8).

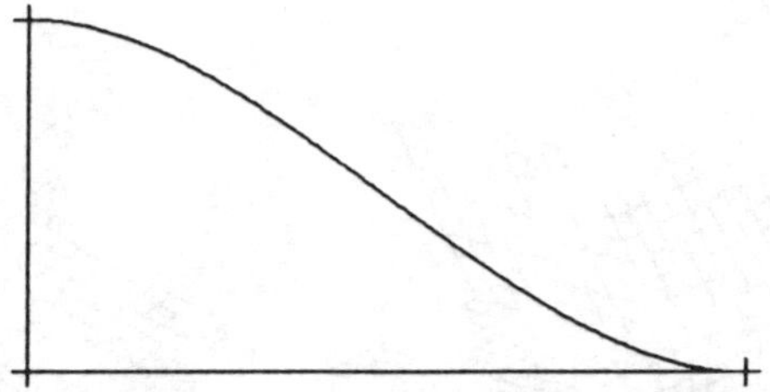

$\textsc{Fig.}$ 10.2. *The weight function* $w_u(t) = 1 - 3t^3 + 2t^2$.

To be able to contruct tangent plane continuity at a corner between the three tensor product surfaces and the box spline surfaces, it is sufficient to ensure equal twist at the corner point of the four surfaces. Since the box spline surface is fixed, this implies that the tensor product surfaces are to be constructed such that the twist vectors of the surfaces become equal to the twist of the box spline surface.

We have the derivative along the boundary of the box spline surface given by

$$(10.5) \qquad \frac{\partial h(t)}{\partial t} = w'_{\mathbf{u}}(t)\lambda_{\mathbf{u}}(t) + w_{\mathbf{u}}(t)\lambda'_{\mathbf{u}}(t) + w'_{\mathbf{v}}(t)\lambda_{\mathbf{v}}(t) + w_{\mathbf{v}}(t)\lambda'_{\mathbf{v}}(t).$$

The mixed partial derivatives of the box spline surface at each end point of the edge are $\lambda'_{\mathbf{u}}(0)$ and $\lambda'_{\mathbf{v}}(1)$, respectively. Thus, evaluation at the end points $t = 0$ and $t = 1$ of (1.5) should yield $\lambda'_{\mathbf{u}}(0)$ and $\lambda'_{\mathbf{v}}(1)$, respectively. To obtain this construction it is necessary that the weight functions $w_{\mathbf{u}}(t)$ and $w_{\mathbf{v}}(t)$ have zero derivatives at both $t = 0$ and $t = 1$. Moreover, $w_{\mathbf{u}}(t)$ and $w_{\mathbf{v}}(t)$ have to be such that $w_{\mathbf{u}}(0) = 1$, $w_{\mathbf{u}}(1) = 0$, $w_{\mathbf{v}}(0) = 0$ and $w_{\mathbf{v}}(1) = 1$.

Let $w_{\mathbf{u}}(t) = 1 - 3t^3 + 2t^2$, the cubic polynomial shown in Fig. 10.2, and $w_{\mathbf{v}}(t) = 1 - w_{\mathbf{u}}(t)$. Then the twist of the surrounding surfaces at each corner of the box spline surface becomes equal to the twist of the box spline surface at the corner. To accomplish this we have to apply tensor product surfaces of degree $|\mathbf{m}|$ to surround a degree $|\mathbf{m}| - 2$ box spline surface to obtain an overall tangent plane continuous surface. Thus, to surround a $C^2$-quartic box spline surface we must use degree 6 tensor product B-spline surfaces, since $\lambda_{\mathbf{u}}$, $\lambda_{\mathbf{v}}$, $w_{\mathbf{u}}$, and $w_{\mathbf{v}}$ are cubic polynomials in $t$.

Another way to accomplish this is to define $w_{\mathbf{u}}$ and $w_{\mathbf{v}}$ as piecewise polynomials or B-splines functions with zero derivatives at $t = 0$ and $t = 1$. With a convenient choice of interior knots according to the knot vector for the cross-boundary derivatives $\lambda_{\mathbf{u}}$ and $\lambda_{\mathbf{v}}$, we are able to define piecewise linear or piecewise quadratic weight functions $w_{\mathbf{u}}$ and $w_{\mathbf{v}}$ with the required end conditions. In the case when $w_{\mathbf{u}}$ and $w_{\mathbf{v}}$ are piecewise quadratic we need degree 5 tensor product B-spline surfaces to completely surround a $C^2$-quartic box spline surface.

To be able to perform these computations we need degree raising algorithms and procedures for the computations of products of B-splines [11].

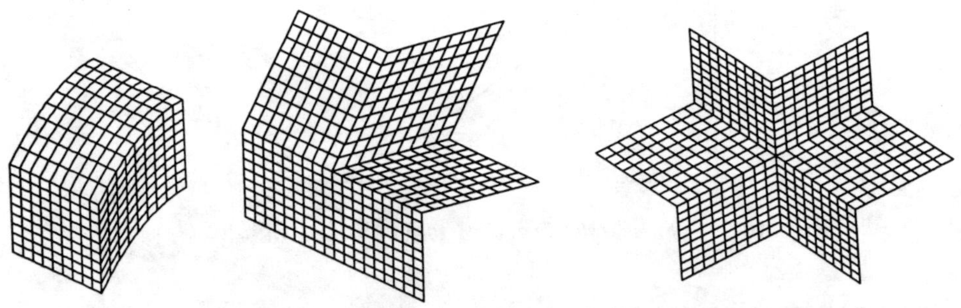

FIG. 10.3. A three-, five-, and six-sided vertex region.

## 10.4. Vertex Regions

In geometric modeling a frequently arising situation is the problem of constructing smooth blends in regions where a number of surfaces meet in one single point. We will refer to neighbourhoods of such regions as vertex regions. The problem discussed here are also covered in [12]. Typical vertex regions are shown in Fig. 10.3.

A construction of fillets and blends for the generation of smooth, nice-looking surfaces at vertex regions will often give rise to the design of surfaces defined over nonrectangular domains. We will focus on such problems using box spline surfaces defined over three-, five-, and six-sided domains. Several authors have studied the problem of generating nonrectangular surfaces for CAGD (see [25] and references therein).

The type of box spline surfaces that are to be constructed depends on the number of surfaces that meet at the vertex, and how the fillets and blends will be constructed in accordance with surrounding surfaces. The construction will consist of a nonrectangular box spline surface at the vertex, surrounded by tensor product B-spline surfaces extending the box spline surface along the edges connected to the vertex. We seek to find an overall tangent plane continuous surface at the vertex region.

As pointed out in the previous section we are able to generate a smoothly blended region by constructing the box spline surface before the surrounding tensor product surfaces. The procedure for the generation of a tangent plane continuous blend at the vertex region is divided into three main steps.

- The first step is to decide the size of the nonrectangular blend at the vertex and the size of the tensor product fillets along the edges connected to the vertex. This is important from a modeler's point of view, but is not considered here.

- The next step is to construct the nonrectangular box spline surface, which is the topic of this section.

- Finally, the box spline surface at the vertex region is surrounded by tensor product B-spline surfaces.

Using box spline $B^{\mathbf{m}}$, with $\mathbf{m} = (m_1, m_2, m_3)$, $m_i \geq 2$, we find that all second-order mixed partial derivatives are uniquely defined at all points in $\mathbb{R}^2$. This makes it possible to approximate arbitrary twist vectors given at each corner of a domain $\Omega$.

Let us now derive the details in the construction of a smooth five-sided vertex region using the $C^2$-quartic box spline $B^{(2,2,2)}$. We want a tangent continuous blend consisting of a five-sided box spline surface and five tensor product surfaces extending the box spline surface along each edge connected to the vertex. Figure 10.4 shows the five-sided parameter domain $\Omega$ given by the parameters $(n_1, n_2, \ell_1, \ell_2) = (6, 6, 0, 3)$ and $\nu = 1$ (see also Fig. 10.1). The circles in the figure indicate the lower left corner of the translates of the box spline $B^{(2,2,2)}$, which are nonzero somewhere on $\Omega$. The filled circles indicate the lower left corner of the translates, which are nonzero at the corners of the domain $\Omega$. In this case we observe that the choice $(n_1, n_2, k_1, k_2) = (6, 6, 0, 3)$ is the "smallest" five-sided domain $\Omega$, so that the construction of an interpolant

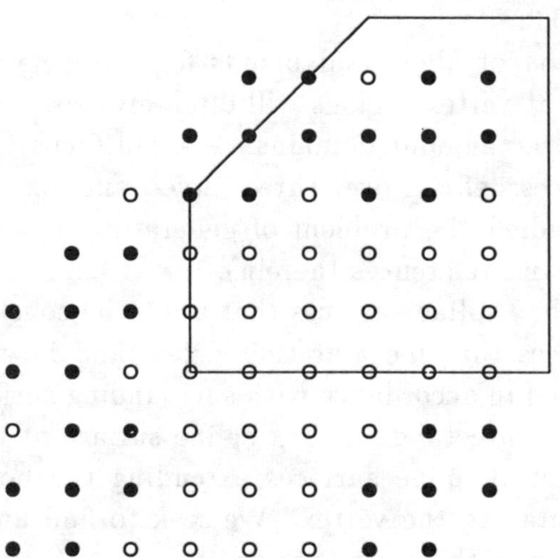

FIG. 10.4. *A five-sided domain $\Omega$ and circles indicating the lower left corner of the support of the $C^2$-quartic box splines $B^{(2,2,2)}$ that overlap $\Omega$. Filled circles indicate the box splines that overlap the corners of $\Omega$.*

at each corner becomes local (see Fig. 10.4).

Let $A_i$, $i = 1,\cdots,5$ in $\mathbb{R}^3$ be the position of the corner points of the surrounding tensor product B-spline surfaces. Moreover, let $B_i$ and $C_i$, $i = 1,\cdots,5$ denote the partial derivatives and $D_i$, $i = 1,\cdots,5$ denote the twist vectors at the points $A_i$, $i = 1,\cdots,5$, respectively. Figure 10.5(a) shows a blow up of the lower left corner of the domain $\Omega$, where the coefficients of the translates of box splines which are nonzero at the corner are indicated by $a_1,\cdots,a_7$. (Note that the position of $a_1,\cdots,a_7$ in Fig. 10.5(a) have nothing to do with the actual position of the coefficients $a_1,\cdots,a_7$ in $\mathbb{R}^3$.)

By evaluation of the $C^2$-quartic box spline $B^{(2,2,2)}$ and its derivatives at the corner we obtain the four equations,

$$a_1 + a_2 + a_3 + 6a_4 + a_5 + a_6 + a_7 = 12A_i,$$
$$-a_1 + a_2 - 3a_3 + 3a_5 - a_6 + a_7 = 8B_i,$$
$$-a_1 - 3a_2 + a_3 - a_5 + 3a_6 + a_7 = 8C_i,$$
$$a_1 - a_2 - a_3 - 2a_4 - a_5 - a_6 + a_7 = 2D_i.$$

With seven unknown and four interpolation conditions we can allow additional assumptions on the coefficients $a_1,\cdots,a_7$. With three additional symmetry

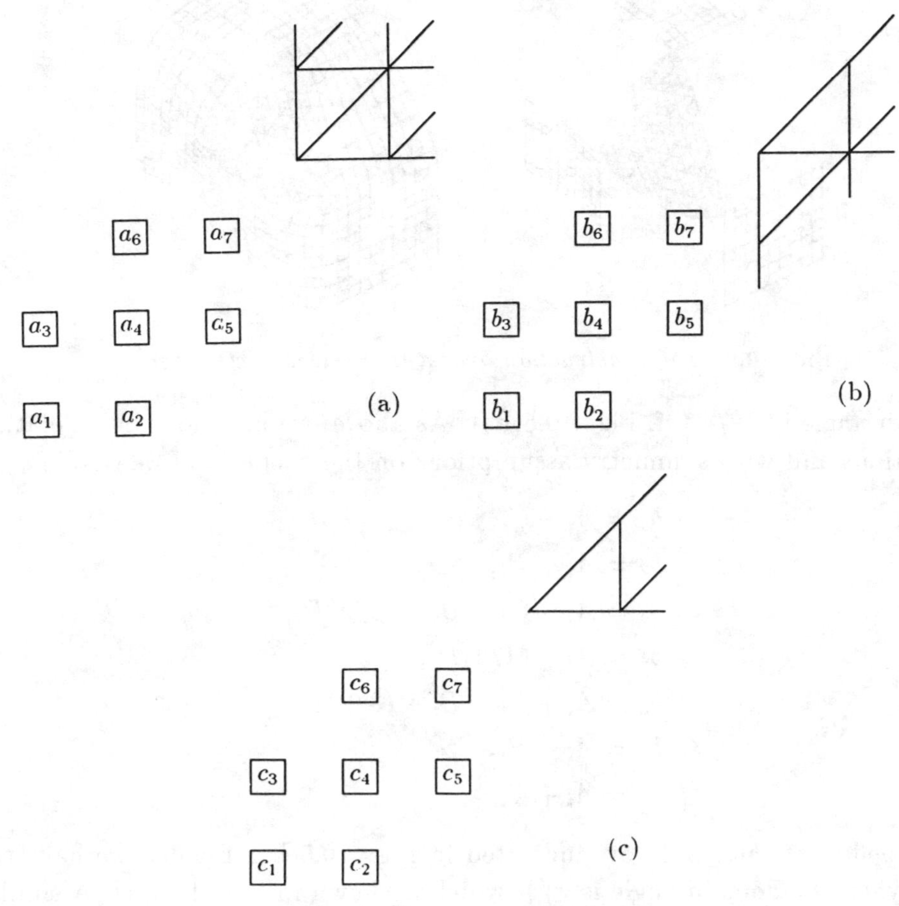

$$\text{F}\small\text{IG. 10.5. } \textit{Three possible domain angles.}$$

assumptions on the coefficients we obtain the following nice configuration:

$$
\begin{aligned}
a_1 &= A_i - B_i - C_i + (3/2)D_i, \\
a_2 &= A_i - C_i, \\
a_3 &= A_i - B_i, \\
a_4 &= A_i - (1/2)D_i, \\
a_5 &= A_i + B_i, \\
a_6 &= A_i + C_i, \\
a_7 &= A_i + B_i + C_i + (3/2)D_i.
\end{aligned}
$$

(10.6)

There exist three types of constructions in the case when $\Omega$ is a convex domain bounded by lines in the three-direction grid. These three situations are shown in Fig. 10.5(a)–(c). All other cases can be obtained by constructions that are symmetric with one of these three. A solution when the domain angle at the corner is $\pi/2$ is given in (10.6). Next, let us consider the case when the

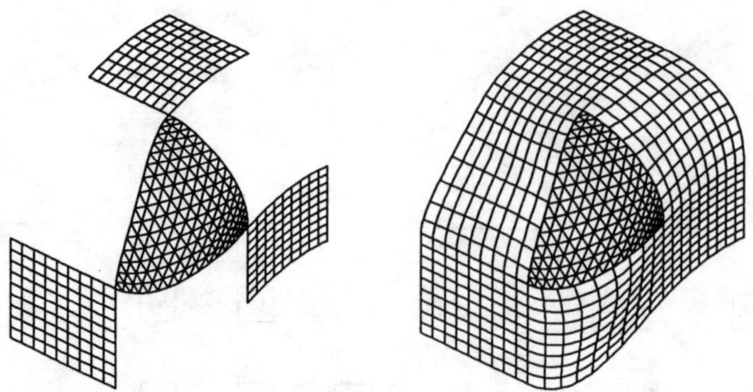

FIG. 10.6. *The construction over a three-sided vertex region.*

domain angle is $3\pi/4$ (cf. Fig. 10.5(b)). As above, we have four interpolation
conditions and with symmetry assumptions on the coefficients we obtain

$$b_1 = A_i - C_i,$$
$$b_2 = A_i + B_i,$$
$$b_3 = A_i - C_i - B_i + (3/2)D_i,$$
$$(10.7) \qquad b_4 = A_i - (1/2)D_i,$$
$$b_5 = A_i + C_i + B_i + (3/2)D_i,$$
$$b_6 = A_i - B_i,$$
$$b_7 = A_i + C_i.$$

The coefficients $b_1, \cdots, b_7$ are indicated in Fig. 10.5(b). Finally, we have the
case when the domain angle is $\pi/4$, which is shown in Fig. 10.5(c). A similar
construction gives

$$c_1 = A_i - C_i,$$
$$c_2 = A_i - C_i + B_i - (3/2)D_i,$$
$$c_3 = A_i - B_i,$$
$$(10.8) \qquad c_4 = A_i + (1/2)D_i,$$
$$c_5 = A_i + B_i,$$
$$c_6 = A_i - B_i + C_i - (3/2)D_i,$$
$$c_7 = A_i + C_i.$$

The situation given in (10.8) does not occur for the domain $\Omega$ shown in Fig.
10.4.

We observe that there are several coefficients that are not affected by the
construction at the corners of the surface, and we need methods to deal with
these extra coefficients. For the construction of nice-looking smooth surfaces
we will now consider a method that minimizes second-order differences, while
keeping the coefficients involved at the corner construction fixed.

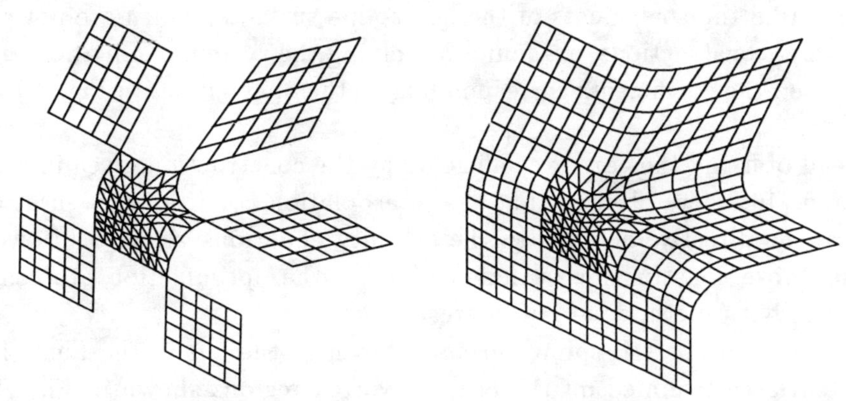

FIG. 10.7. *The construction over a five-sided vertex region.*

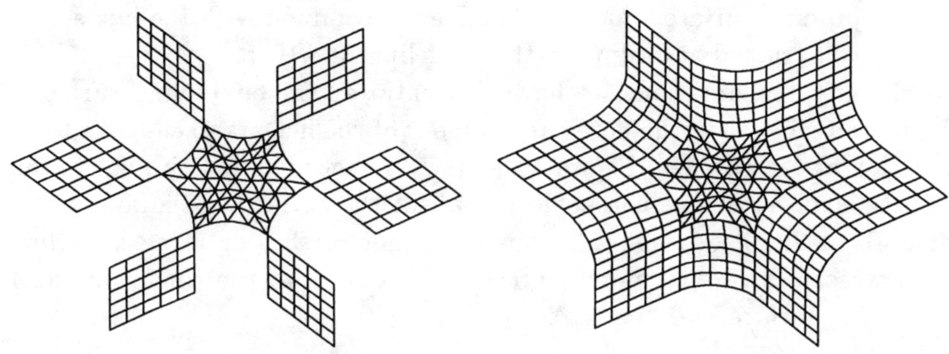

FIG. 10.8. *The construction over a six-sided vertex region.*

Let $\mathbf{c} = (c_1, \cdots, c_n)$ be an ordering of the coefficients corresponding to all translates of the box spline $B^{\mathbf{m}}$ that are nonzero somewhere on $\Omega$. Some of these coefficients are given by (10.6), (10.7), or (10.8). To obtain a nice surface we assign values to the other coefficients by solving the Lagrange equations

$$(10.9) \qquad \begin{pmatrix} Q & A^T \\ A & O \end{pmatrix} \begin{pmatrix} \mathbf{c} \\ \mathbf{y} \end{pmatrix} = \begin{pmatrix} \mathbf{0} \\ \mathbf{b} \end{pmatrix},$$

where $\mathbf{y}$ are the Lagrange multipliers (see [21]). This gives a solution to the following quadratic optimization problem,

$$\text{Minimize } \mathbf{c}^T Q \mathbf{c}$$

$$\text{subject to } A\mathbf{c} = \mathbf{b}.$$

Here $A\mathbf{c} = \mathbf{b}$ denotes the equality constraints, which fix the coefficients according to (10.6), (10.7), or (10.8). Moreover, we have that $Q = U^T V U$, where the matrix $U$ performs second-order differences on the coefficients and where

$$V = \text{diag}(v_1, \cdots, v_n)$$

is a suitable weight matrix. Details on $U$ and $V$ can be found in [1],[14]. In Figs. 10.6–10.8 the coefficients of the box spline surfaces, which are not given by the corner constructions, are found by solving the system of equations given in (10.9), i.e., the coefficients corresponding to the open circles in Fig. 10.4 are found.

Instead of fixing the corner coefficients by the construction given in (10.6), (10.7), and (10.8), we observe that the interpolation conditions given at each corner can be directly included in the equality equations $A\mathbf{c} = \mathbf{b}$. However, since the above construction is independent of the optimization problem we believe that both methods are of interest.

The tensor product B-spline surfaces shown to the left in the Figs. 10.6–10.8 are surfaces trimmed off the original vertex regions shown in Fig. 10.3. We then construct box spline surfaces which match position, partial derivatives and twist vectors of the respective trimmed off surfaces. To the right in Figs. 10.6–10.8 the final tangent plane continuous vertex regions consisting of one box spline surface and a number of tensor product B-spline surfaces are shown. The tensor product surfaces that have an edge in common with the box spline surface are constructed according to the guidelines in §10.4.

To simplify the notation, we have concentrated on box spline surfaces, which have no more than the necessary number of coefficients to ensure a local construction at each corner. Figure 10.4 shows a five-sided domain, where the coefficients involved at one corner do not affect the surface at the four other corners. By defining box spline surfaces on finer meshes or using knot line insertion we can introduce more degrees of freedom into the construction of the box spline surface.

## References

[1] E. Arge and M. Dæhlen, *Grid point interpolation on finite regions by $C^1$ box splines*, J. Comp. Appl. Math, 1992, to appear.

[2] C. de Boor, *A practical guide to splines*, Springer-Verlag, New York, 1978.

[3] C. de Boor and R. DeVore, *Approximations by smooth multivariate splines*, Trans. Amer. Math. Soc., 276 (1983), pp. 775–785.

[4] C. de Boor and K. Höllig, *B-splines from parallelepipeds*, J. Analyse Math., 42 (1982/83), pp. 99–115.

[5] ——, *Bivariate box splines and pp functions on a three-direction mesh*, J. Comput. Appl. Math., 9 (1983), pp. 13–28.

[6] A. S. Cavaretta and C. A. Micchelli, *The design of curves and surfaces by subdivision algorithms*, Mathematical Methods in Computer Aided Geometric Design, T. Lyche and L. L. Schumaker, eds., Academic Press, New York, 1989, pp. 115–153.

[7] C. K. Chui, *Multivariate Splines*, CBMS-NSF Regional Conference Series in Applied Mathematics, Society for Industrial and Applied Mathematics, Philadelphia, 1988.

[8] E. Cohen, T. Lyche, and R. Riesenfeld, *Discrete box-splines and refinement algorithms*, Comput. Aided Geom. Des., 1 (1984), pp. 131–148.

[9] E. Cohen, T. Lyche, and L. L. Schumaker, *Degree raising for splines*, J. Approx. Theory, 46 (1986), pp. 170–181.

[10] M. Dæhlen, *On the evaluation of box splines*, in Mathematical Methods in Computer Aided Geometric Design, T. Lyche and L. L. Schumaker, eds., Academic Press, New York, 1989, pp. 167–179.

[11] M. Dæhlen and T. Lyche, *Bivariate interpolation with quadratic box splines*, Math. Comp., 51 (1988), pp. 219–230.

[12] M. Dæhlen and V. Skytt, in *Modelling non-rectangular surfaces using box splines*, Mathematics of Surfaces III, D. Handscomb, ed., Clarendon Press, Oxford, 1989, pp. 285–300.

[13] M. Dæhlen, *Knotline removal on box spline surfaces*, preprint.

[14] M. Dæhlen and T. Lyche, *Box splines and applications*, Geometric Modelling: Methods and Applications, D. Roller and H. Hagen, eds., Springer-Verlag, New York, 1991, pp. 35–94.

[15] W. Dahmen and C. A. Micchelli, *Recent progress in multivariate splines*, Approximation Theory IV, C. K. Chui, L. L. Schumaker, and J. Ward, eds., Academic Press, New York, 1983, pp. 27–121.

[16] ——, *Subdivision algorithms for generation of of box-spline surfaces*, Comput. Aided Geom. Des., 1 (1984), pp. 115–129.

[17] ——, *On the local linear independence of translates of a box spline*, Studia Math., 82 (1985), pp. 243–262.

[18] T. N. T. Goodman, *Polyhedral splines*, in Computation of Curves and Surfaces, W. Dahmen, M. Gasca, and C. A. Micchelli, eds., Kluwer, Dordrecht, 1990, pp. 347–382.

[19] K. Höllig, *Box-splines surfaces*, in Mathematical Methods in Computer Aided Geometric Design, T. Lyche and L. L. Schumaker, eds., Academic Press, New York, 1989, pp. 385–402.

[20] R-Q. Jia, *Linear independence of translates of box splines*, J. Approx. Theory, 40 (1984), pp. 158–160.

[21] D. G. Luenberger, *Linear and Nonlinear Programming*, Addison-Wesley, London, 1984.

[22] T. I. Mueller, *Geometric Modelling with multivariate B-splines*, Dissertation, Dept. of Comp. Science, University of Utah, 1986.

[23] M. A. Sabin, *The use of piecewise forms for the numerical representation of shapes*, Dissertation, Hungarian National Academy of Sciences, 1977.

[24] L. L. Schumaker, *Spline Functions: Basic Theory*, Wiley & Sons, New York, 1981.

[25] T. Varady, *Survey on new results in N-sided patch generation*, The Mathematics of Surfaces II, R. R. Martin, ed., Oxford Science Pub., Oxford, 1987, pp. 203–235.